Affine Bernstein Problems *and* Monge-Ampère Equations

Affine Bernstein Problems *and* Monge-Ampère Equations

An-Min Li (Sichuan University, China)
Ruiwei Xu (Henan Normal University, China)
Udo Simon (Technische Universität Berlin, Germany)
Fang Jia (Sichuan University, China)

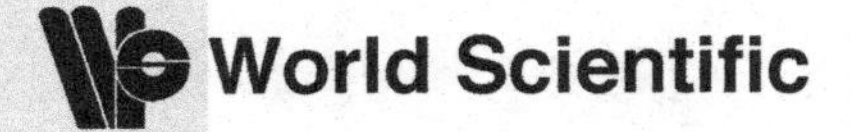

NEW JERSEY • LONDON • SINGAPORE • BEIJING • SHANGHAI • HONG KONG • TAIPEI • CHENNAI

Published by

World Scientific Publishing Co. Pte. Ltd.
5 Toh Tuck Link, Singapore 596224
USA office: 27 Warren Street, Suite 401-402, Hackensack, NJ 07601
UK office: 57 Shelton Street, Covent Garden, London WC2H 9HE

British Library Cataloguing-in-Publication Data
A catalogue record for this book is available from the British Library.

AFFINE BERNSTEIN PROBLEMS AND MONGE-AMPÈRE EQUATIONS

ISBN-13 978-981-281-416-6
ISBN-10 981-281-416-7

Printed in Singapore.

Preface

Many geometric problems in analytic formulation lead to important classes of PDEs. Naturally, since all such equations arise in geometric context, geometric methods play a crucial role in their investigation. A classical example is given by the Euclidean Minkowski problem: the study of hyperovaloids with prescribed Gauß curvature in terms of the Euclidean unit normal field. For the history up to the early 70's see Pogorelov's monograph [77] from 1975 and the paper of Cheng and Yau [25] from 1976. The study of Minkowski's problem and the related regularity was essential for the understanding of certain Monge-Ampère type equations on the Euclidean sphere.

Our monograph is devoted to the interplay of global differential geometry and PDEs, more precisely to the study of some types of non-linear higher order PDEs; most of them have their origin in the affine hypersurface theories. Particular examples include the PDEs defining affine spheres and affine maximal hypersurfaces, resp., and the constant affine mean curvature equation.
Wide use of geometric methods in studying PDEs of affine differential geometry was initiated by E. Calabi and continued by A.V. Pogorelov, S.Y. Cheng-S.T. Yau, A.-M. Li, and, during the last decade, e.g. by N.S. Trudinger-X.J. Wang, A.-M. Li's school, and other authors.
The contributions of E. Calabi and S.Y. Cheng-S.T. Yau had a particularly deep influence on the development of this subject. According to the foreword in [25] this paper originated from discussions with E. Calabi and L. Nirenberg and results of both on the same topic; for further historical details and references we refer to [19], [20], [58], [76].
In problems involving PDEs of Monge-Ampère type it is often the case that the unknown solution is a convex function defining locally a nonparametric hypersurface for which it is possible to choose a suitable relative normalization and investigate the induced geometry. We refer to this process as *geometric modelling.* The choice of the normalization can be described in a unified and systematic manner in the context of relative hypersurface theory; for this theory see [58], [87], [88].

The next step involves derivation of estimates of various geometric invariants; a correct choice of a normalization is very important for successful completion of this step. Ultimately, such estimates are crucial for proving the existence and uniqueness, respectively, of solutions to the PDE.

In chapter 1 we start with a summary of basic tools; very good sources for that are the monographs [37], [50] and [58]. For a better understanding of the modelling techniques, in chapters 2 and 3 the authors give a selfcontained summary of relative hypersurface theory. Moreover, for the global study, we consider different notions of completeness in sections 4.2 and 5.9.

Chapters 4-6 are the central part of the monograph. They contain important PDEs from affine hypersurface theory: the PDEs for affine spheres, affine maximal surfaces, and constant affine mean curvature hypersurfaces. The PDE for improper affine spheres over $\mathbb{R}^2$ first was studied by Jörgens in the paper [49]; Calabi [19] extended the result to the dimensions $n \leq 5$, and finally Pogorelov to any dimension [76]. Later, Cheng and Yau extended Pogorelov's version and gave a simpler and more analytic proof in [25]; concerning this paper and Calabi's influence, see our remarks above. Nowadays, in the literature the Theorem is cited as *Theorem of Jörgens-Calabi-Pogorelov.* In section 4.4 we present the geometric Calabi-Cheng-Yau proof for this theorem, [19], [25]. Afterwards we study a generalization of this theorem. As the proof of the generalization is relatively simple in dimensions $n \leq 4$, we use both proofs for a comparison of the geometric modelling procedure:

(i) In the proof of the Theorem of Jörgens-Calabi-Pogorelov we use Blaschke's normalization.

(ii) In the second example we give a proof of the generalization. Now we use a constant normalization of a graph and its induced geometry; to our knowledge it was first used by Calabi within this context.

Sections 4.5.5 and 4.6.2 present such comparisons of proofs with different modelling, emphasizing the interplay between the geometric model chosen and the PDE considered. In arbitrary dimension the proof of the extension of the Theorem of Jörgens-Calabi-Pogorelov is complicated, thus we carefully structure the proof as guideline for the reader (see section 4.5.7).

In chapter 5 we derive the Euler-Lagrange equation of affine maximal hypersurfaces. The topic of this chapter is given by different versions of the so called "Affine Bernstein Problem", in particular the "Affine Bernstein Conjectures" in dimension $n = 2$. They are due to Chern and Calabi, resp., and were solved during the last decade. In 2000, Trudinger and Wang solved Chern's conjecture in dimension $n = 2$ [91]; later, Li and Jia [52], and also Trudinger and Wang [92], solved Calabi's conjecture for two dimensions independently, using quite different methods. In section 5.7 we treat *Calabi's Affine Bernstein Problem* in dimensions $n = 2$ and $n = 3$.

The final chapter studies constant affine mean curvature hypersurfaces. In dimension $n = 2$ the problem was solved in case the constant is positive; in case the constant is zero we have again the "Affine Bernstein Problems". The case of negative constant mean curvature has been solved partially only, so far. For any bounded convex domain, we can construct a Euclidean complete affine hypersurface with negative constant affine mean curvature solving a boundary value problem for a fourth order PDE.

The monographs [5] and [37] give a good basis for the geometric theory of Monge-Ampère equations. Our monograph gives a geometric method for the study of Monge-Ampère equations and fourth order nonlinear PDEs arising in affine differential geometry. There are recent related papers from A.-M. Li's school (e.g. [24]), and there are extensions to Kähler geometry and projective Blaschke manifolds [63]. Other interesting results concern global affine maximal surfaces with singularities, see e.g. [2], [3], [4], [34], [69].

The authors present three generations of geometers. U. Simon finished his doctoral thesis with K.P. Grotemeyer at the FU Berlin in 1965, and from his lectures he became interested in global differential geometry. U. Simon became a professor of mathematics at TU Berlin in 1970. A.-M. Li started his studies at Peking University in 1963, but because of the cultural revolution he could not finish his MS before 1982. Following a recommendation of S.S. Chern, he came as AvH fellow to the TU Berlin in 1986 the first time, and there he finished his doctoral examination with U. Simon, U. Pinkall and K. Nomizu. A.-M. Li has been a professor of mathematics at Sichuan University since 1986, successfully guiding research groups since then. A.-M. Li was also the advisor of F. Jia (PhD 1997) and R. Xu (PhD 2008) at Sichuan University, both are now professors themselves, F. Jia at Sichuan University (1997), R. Xu at Henan Normal University since 2008.
The homepages of our Chinese-German cooperation give some more details, for the momentary project see http://www.math.tu-berlin.de/geometrie/gpspde/.
Blaschke's interest in the global study of submanifolds was important for Chern's decision to go to Hamburg in 1934, and not to Göttingen. Their interest in global problems influenced the following generations. We aim to stimulate young geometers again.

Acknowledgements. The authors thank the following institutions for financial support that made possible joint work on the topic at Chengdu and Berlin, respectively: Alexander von Humboldt Stiftung (AvH), Deutsche Forschungsgemeinschaft (DFG), Dierks von Zweck Stiftung Germany, NSF China (10631050, 10926172, 10871136), RFDP, Sichuan University, TU Berlin and Henan Normal University. Moreover, TU Berlin made it possible that the authors could work together on this monograph in Berlin for several months in 2008 and 2009.

We thank Mr. Min Xiong, Sichuan University, for a careful reading of our manuscript, and Vladimir Oliker from Emory University for very helpful discussions on the topic.

A.-M. Li, R. Xu, U. Simon, F. Jia

January 2010

Contents

Chapter 1

Basic Tools

1.1 Differentiable Manifolds

1.1.1 *Manifolds, connections and exterior calculus*

We denote by M a connected *differentiable manifold* of *dimension* $n \geq 2$. At a point $p \in M$ we denote the *tangent space* by T_pM and its dual by T_p^*M, accordingly the *tangent bundle* by TM and the *cotangent bundle* by T^*M. As far as there is no emphasis on the degree of differentiability the term *"differentiable"* means C^∞; as usual we write $f \in C^\infty(M)$ when f is a C^∞-function on M. We denote vectors and vector fields by $v, w, ...$ and the space of vector fields by $\mathfrak{X}(M)$.

Connections. We denote an *affine connection* by ∇, and use this symbol also to indicate *covariant differentiation* in terms of ∇ in case we are using the invariant calculus. All connections considered are torsion free.
The covariant differentiation of a one-form η is defined by:

$$(\nabla_v \eta)(w) := v(\eta(w)) - \eta(\nabla_v w).$$

Exterior calculus. An alternating $(0, r)$-tensor field on M is called an *exterior differential form of degree* r, or simply an r*-form.* Denote by $\Lambda^r(M)$ the set of all smooth exterior differential forms of degree r and define

$$\Lambda(M) := \Lambda^0(M) \oplus \Lambda^1(M) \oplus ... \oplus \Lambda^n(M),$$

where $\Lambda^0(M) := C^\infty(M)$. With respect to exterior multiplication $\wedge$ the set $\Lambda(M)$ is an associative algebra, called the *exterior algebra* on M.
It is well known that there is a unique linear map $d : \Lambda(M) \rightarrow \Lambda(M)$, called the *exterior differentiation*, that satisfies the following rules:
(i) $d : \Lambda^r(M) \rightarrow \Lambda^{r+1}(M)$,
(ii) $d(f) := df$ for $f \in C^\infty(M)$,
(iii) for $\alpha \in \Lambda^r(M)$ and $\beta \in \Lambda^k(M)$ we have: $d(\alpha \wedge \beta) = d\alpha \wedge \beta + (-1)^r \alpha \wedge d\beta$,
(iv) $d \cdot d = 0$.

The *exterior derivative* and the covariant derivative of η are related by:

$$d\eta(v, w) = (\nabla_v \eta)(w) - (\nabla_w \eta)v.$$

In affine hypersurface theory there appear different affine connections, in such cases we use additional marks. For $f \in C^\infty(M)$, we write $Hess_\nabla f$ for the ∇-*covariant Hessian.*

Cartan's Lemma. *Let $\{\omega^1, ..., \omega^r\}$ be a system of linearly independent 1-forms for $1 \leq r \leq n$ and $\{\eta^1, ..., \eta^r\}$ be another system of 1-forms satisfying*

$$\sum_{s=1}^{r} \omega^s \wedge \eta^s = 0.$$

Then

$$\eta^s = \sum_{p=1}^{r} c_p^s \, \omega^p$$

with symmetric coefficients c_p^s.

Cartan's moving frames. Let $O \subset M$ be an open set and $\{e_1, ..., e_n\}$ differentiable vector fields on O which are pointwise linearly independent. We call $\{e_1, ..., e_n\}$ a *moving frame* on O. Via duality there are linearly independent, differentiable one-forms $\{\omega^1, ..., \omega^n\}$.
For any tangent vector v in TO one has

$$\nabla_v e_j = \sum_k \omega_j^k(v) \, e_k.$$

For $v = e_i$ one usually adapts a notation from the so called local calculus (see below) and writes:

$$\omega_j^k(e_i) := \Gamma_{ij}^k;$$

one calls the coefficients Γ_{ij}^k *Christoffel symbols.* The coefficients ω_j^k are linear in v; thus the collection $\{ \omega_j^k \mid j, k = 1, ..., n\}$ forms a matrix of differentiable one-forms; they are called *connection one-forms.*
The connection one-forms appear again in the *first Cartan structure equations*, giving the *exterior derivative* of ω^i:

$$d\omega^i = \sum_j \omega^j \wedge \omega_j^i.$$

Curvature. For a given connection ∇ consider the *curvature tensor* $R := R_\nabla$:

$$R(v, w)z := \nabla_v \nabla_w z - \nabla_w \nabla_v z - \nabla_{[v,w]} z.$$

This definition follows the sign-convention in [50]. For fixed tangent vectors v, w one considers $R(v, w) : z \mapsto R(v, w)z$ as linear operator, called *curvature operator.* Taking the trace *tr* of this linear map, we get a (0,2)-tensor field, the *Ricci tensor,* denoted by *Ric*:

$$Ric(v, w) : tr\{z \mapsto R(z, v)w\}.$$

It is symmetric if and only if the connection locally admits a parallel volume form; this volume form is unique modulo a non-zero constant factor.

Define the 2-form $\Omega^i_j(v,w) := \omega^i(R(v,w)e_j)$, then *Cartan's second structure equations* read

$$d\omega^i_j = \sum_k \omega^k_j \wedge \omega^i_k + \Omega^i_j.$$

Local notation. Consider a local Gauß basis $\{\partial_1, ..., \partial_n\}$ associated to local coordinates $\{x^1, ..., x^n\}$. As usual we write the dual one-forms as $\{dx^1, ..., dx^n\}$. Using local coordinates, it is convenient to denote a point with coordinates $\{x^1, ..., x^n\}$ just by x.
A connection ∇ locally is uniquely determined by its coordinate-components Γ^k_{ij}, called *Christoffel symbols*, implicitly defined by:

$$\nabla_{\partial_i}\partial_j = \Gamma^k_{ij}\,\partial_k.$$

A connection ∇ is torsion free if and only if the Christoffel symbols satisfy the symmetry relation $\Gamma^k_{ij} = \Gamma^k_{ji}$. As already stated, we consider torsion free connections only.
Concerning the curvature tensor, we write $R(\partial_i, \partial_j)\partial_k =: R^h{}_{kij}\,\partial_h$ and by contraction for the Ricci tensor $Ric(\partial_i, \partial_k) = R^h{}_{ihk} =: R_{ik}$.
In a local coordinate system, we denote *partial derivatives* of $f \in C^\infty(M)$ by

$$f_i = \partial_i f, \quad f_{ij} = \partial_i\partial_j f, \qquad \text{etc.},$$

while we denote *covariant derivatives* in terms of a given connection by

$$f_{,i}\,, \qquad f_{,ij}\,, \qquad \text{etc.}$$

Bianchi identities. The curvature tensor satisfies two cyclic identities; in local notation, for torsion free connections, they have the following form:

$$R^i{}_{jkl} + R^i{}_{klj} + R^i{}_{ljk} = 0,$$
$$R^i{}_{jkl,m} + R^i{}_{jlm,k} + R^i{}_{jmk,l} = 0.$$

Ricci identities. Higher order covariant derivatives do not commute in general; their difference depends on the curvature of the connection. We will apply this in case of a torsion free connection. Let T be an (r,s)-tensor field. We write the *Ricci identities* in local notation:

$$T^{j_1\ldots j_r}_{i_1\ldots i_s\,,kl} - T^{j_1\ldots j_r}_{i_1\ldots i_s\,,lk} = \sum_{q=1}^{s} T^{j_1\ldots j_r}_{i_1\ldots i_{q-1}hi_{q+1}\ldots i_s} R^h{}_{i_qkl} - \sum_{p=1}^{r} T^{j_1\ldots j_{p-1}hj_{p+1}\ldots j_r}_{i_1\ldots i_s} R^{j_p}{}_{hkl}\,.$$

The covariant Hessian. For $f \in C^\infty(M)$ and a given torsion free connection ∇ the *covariant Hessian* is defined by

$$(Hess_\nabla f)(v,w) := v(wf) - (\nabla_v w)f.$$

As ∇ is torsion free, the (0,2)-field $Hess_\nabla f$ is symmetric.

1.1.2 *Riemannian manifolds*

A manifold M together with a differentiable, symmetric, positive definite 2-form g on M is a *Riemannian manifold*, in short notation (M, g). The *metric tensor* g, in short *metric*, induces the following structures: A *distance function* $d : M \times M \to \mathbb{R}$, thus (M, d) is a metric space; on each tangent space one has an *inner* or *scalar product*, again denoted by g; a *norm* on r-forms, denoted by $\|A\|_g$ for an r-form A; and the *Riemannian volume form* $dV := dV(g)$.

Fundamental Theorem and Ricci Lemma. *There is exactly one torsion free connection on M, denoted by $\nabla(g)$, that is compatible with the metric g, which means:*

$$0 = \nabla(g)_k \, g_{ij} = \partial_k g_{ij} - \Gamma^h_{kj} \, g_{ih} - \Gamma^h_{ik} \, g_{hj},$$

or in Cartan's notation

$$0 = dg_{ij} - \sum g_{ik} \, \omega^k_j - \sum g_{kj} \, \omega^k_i.$$

This connection is called the *Levi-Civita connection* of g, the compatibility condition is called the *Ricci Lemma*. The Ricci Lemma expresses the fact that the metric g is *parallel* with respect to the Levi-Civita connection: $\nabla(g) \; g = 0$. It follows from the Ricci Lemma that $\nabla(g)$ is completely determined by g.

Curvature. Following the sign-convention from above, the Levi-Civita connection defines the *curvature tensor* $R(g)$ as (1,3)-tensor field and its symmetric *Ricci tensor* $Ric(g)$. Contraction by the metric gives the *normed scalar curvature* κ, defined by $n(n-1)\kappa := tr_g \, Ric(g)$.
If there is no risk of confusion we will skip the mark g and simply write ω, ∇, R, ..., $\|A\|$ etc; moreover, if the context is clear, we will also write $R = R(g)$ for the *Riemannian curvature tensor* which is a (0,4)-form.
The metric defines a conformal Riemannian class, and for $n \geq 3$ the simplest invariant of this class is the *Weyl conformal curvature tensor* W:

$$\begin{aligned}(n-2)W(u,v)w :=& (n-2)R(u,v)w - n\kappa(g(v,w)u - g(u,w)v) \\ &- [Ric(u,w)v - Ric(v,w)u + Ric^\sharp(v)g(u,w) - Ric^\sharp(u)g(v,w)].\end{aligned}$$

Here $Ric^\sharp$ is the *g-associated Ricci operator*. It is well known that the Riemannian curvature tensor is an algebraic curvature tensor, see [36]. It has an orthogonal decomposition into 3 irreducible components with respect to the orthogonal group associated to g; see pp. 45-49 in [6]. One component, namely the conformal curvature tensor, is totally traceless; the second is Ricci-flat; the third one looks - modulo a constant non-zero factor - like a curvature tensor of constant curvature.

Orthonormal frames. On a Riemannian manifold (M, g) one often picks frames $\{e_1, ..., e_n\}$ to be orthonormal at every point of an open set O. Then
$\omega^k_j(v) = g(\nabla_v e_j, e_k)$, which implies

$$\omega^k_j + \omega^j_k = 0 \quad \text{and} \quad \Omega^k_j + \Omega^j_k = 0.$$

Local notation. With respect to a Gauß basis or a frame, the matrix associated to g usually is written (g_{ij}), and its inverse matrix by (g^{ik}), thus the coefficients satisfy $g_{ij}g^{ik} = \delta_j^k$. As usual the operations of lowering and raising indices via the metric g are defined; obey the Einstein summation convention.
For the local notation of derivatives we refer to the notational convention above; in the Riemannian case, for covariant derivatives, we use the Levi-Civita connection; all exceptions will be explicitly stated.

The Laplacian. For $f \in C^\infty(M)$, we write $Hess_g f$ for the *covariant Hessian* in terms of the Levi-Civita connection, its *trace* with respect to g, denoted by tr_g, defines the *Laplace operator*:

$$\Delta f := tr_g \, Hess_g f.$$

In terms of a local representation of the metric g, the Laplacian reads:

$$\Delta = \frac{1}{\sqrt{\det(g_{kl})}} \sum \frac{\partial}{\partial x^i} \left(g^{ij} \sqrt{\det(g_{kl})} \, \frac{\partial}{\partial x^j} \right).$$

1.1.3 *Curvature inequalities*

Lemma. *Let (M, g) be an n-dimensional Riemannian manifold. We consider the Riemannian curvature tensor R, the Ricci tensor Ric and the normed scalar curvature κ. Then we have the inequalities:*

$$\|Ric\|^2 \geq n(n-1)^2 \kappa^2, \tag{1.1.1}$$

$$\|R\|^2 \geq \tfrac{2}{n-1} \|Ric\|^2, \tag{1.1.2}$$

$$\|R\|^2 \geq 2n(n-1)\kappa^2. \tag{1.1.3}$$

Equality in the first relation holds if and only if (M, g) is Einstein. The second equality holds if and only if (M, g) is conformally flat, the third if and only if (M, g) has constant sectional curvature.

Proof. Proofs of this type of inequalities are standard. To prove the first inequality, calculate the squared norm of the traceless part of the Ricci tensor:

$$0 \leq \|Ric - (n-1)\kappa \, g\|^2.$$

To prove the second inequality, consider the Weyl conformal curvature tensor W from section 1.1.2 above and calculate $0 \leq \|R - W\|^2$. The third inequality is a combination of the two foregoing inequalities. For the discussion of equality in this case recall that a conformally flat Einstein space is of constant sectional curvature. ∎

Inequalities for r-forms. As far as we know inequalities of the above type were used by E. Calabi the first time. Let (M, g) be a Riemannian manifold.
1. In case an r-form satisfies symmetries and skew-symmetries like arbitrary curvature tensors, the above examples indicate how to prove optimal inequalities.

2. The following sketches a simple method to prove optimal inequalities for arbitrary r-forms on (M, g); see [75]. Let D be an r-form for $r \geq 2$. Let $\sigma(D)$ be the normed, totally symmetrized tensor coming from D:

$$\sigma(D)_{i_1 \dots i_r} := \frac{1}{r!} \sum_{\sigma} D_{\sigma(i_1) \dots \sigma(i_r)};$$

here the summation runs over all permutations σ of the r-tuple. Let $\widetilde{D}$ be the traceless part of $\sigma(D)$ with respect to the metric g. Then

$$0 \leq \|\widetilde{D}\|^2 \leq \|D\|^2.$$

Equality on the right holds if and only if D itself is totally symmetric and traceless.

1.1.4 *Geodesic balls and level sets*

Let $\Omega \subset \mathbb{R}^n$ be a domain; a function $f : \Omega \rightarrow \mathbb{R}$ is called *convex* if, for all $0 \leq t \leq 1$ and $x, y \in \Omega$ such that $tx + (1-t)y \in \Omega$, we have

$$f(tx + (1-t)y) \leq tf(x) + (1-t)f(y).$$

Let f be a smooth convex function defined on $\mathbb{R}^n$. Given a constant $C > 0$ and $\ell(x) = f(x_0) + (\text{grad } f)(x_0) \cdot (x - x_0)$ a supporting hyperplane to f at $(x_0, f(x_0))$, a *section* of f at height C is the *level set*

$$S_f(x_0, C) := \{x \in \mathbb{R}^n \mid f(x) < \ell(x) + C\}.$$

In particular, if we neglect the point where f attains its minimum, we use a shorter notation to denote the level set

$$S_f(C) := \{x \in \mathbb{R}^n \mid f(x) < C\}.$$

This set is convex. We remark that in case the convex function f is defined only on a convex open set $\Omega \subset \mathbb{R}^n$, the sections of f at $x_0 \in \Omega$ mean the sets $\overline{S_f(x_0, C)} \subset \Omega$. Denote by $\mathcal{S}(\Omega, C)$ the class of strictly convex C^∞-functions f, defined on Ω, such that

$$\inf_{\Omega} f(x) = 0, \qquad f(x) = C \qquad \text{on} \qquad \partial\Omega.$$

$B_R(p)$ denotes the open Euclidean ball with center at p and with radius R.
$B_a(p, G)$ denotes the open geodesic ball with respect to the metric G, centered at p with radius a.
$\|\cdot\|_G$ denotes the norm of a vector or a tensor with respect to the Riemann metric G, while $\|\cdot\|_E$ denotes the norm of a vector with respect to the canonical Euclidean metric.

1.2 Completeness and Maximum Principles

1.2.1 *Topology and curvature*

We list some results about completeness in a form that we will need. For the first three theorems, see [33]. Standard references for maximum principles are [35] and [78].

Theorem. (H. Hopf - W. Rinow). *For a Riemannian manifold (M, g) the following conditions are equivalent:*
(i) *(M, d) is a complete metric space;*
(ii) *(M, ∇) is geodesically complete;*
(iii) *every topologically closed and bounded subset is compact.*

Theorem. (J. Hadamard - E. Cartan). *Let (M, g) be complete with non-positive sectional curvature. Then, for every $p \in M$, the exponential map is a covering map. In particular, if M is simply connected then M is diffeomorphic to $\mathbb{R}^n$.*

The following theorem originates from a result of Hadamard for compact surfaces without boundary and was extended in several steps to a very general result [100]; we need the following part of it.

Theorem. (J. Hadamard - R. Sacksteder - H. Wu). *Let (M, g) be an n-dimensional complete, noncompact, orientiable hypersurface in $\mathbb{R}^{n+1}$ with positive sectional curvature. Then there exists $p \in M$ such that M can be represented as graph of a non-negative, strictly convex function over the tangent plane $T_pM \subset \mathbb{R}^{n+1}$.*

Theorem. (S.B. Myers). *Let (M, g) be complete with Ricci curvature positively bounded from below:*

$$Ric \geq (n-1) \cdot c^2 g$$

where $0 < c \in \mathbb{R}$. Then the diameter satisfies $diam\,(M, g) \leq diam\,(S^n(\frac{1}{c}))$, where $\frac{1}{c}$ is the radius. In particular, M is compact with finite fundamental group.

1.2.2 *Maximum principles*

Maximum principle. (E. Hopf). *In a bounded domain $\Omega \subset \mathbb{R}^n$, let us consider a second order differential operator of the form*

$$L = \sum_{i,j} a_{ij}(x)\, \partial_i\, \partial_j + \sum_i b_i(x)\, \partial_i$$

with continuous, symmetric, positive definite coefficient matrix $(a_{ij}(x))$, continuous functions b_i and $x \in \Omega$. Assume that the differentiable function $f : \Omega \to \mathbb{R}$ satisfies the conditions
(i) *$L\,f \geq 0$ in Ω;*
(ii) *there is a point $x_0 \in \Omega$ such that $f(x) \leq f(x_0)$ for all $x \in \Omega$.*
Then f is constant in Ω: $f(x) = f(x_0)$.

Remark. **(i)** Of course, one can reverse all inequalities; then the assertion holds true.
(ii) Trivially, the Laplacian is a special case of an elliptic operator.

Harmonic functions. (S.T. Yau [104]). *Let (M, g) be a complete, non-compact Riemannian n-manifold with non-negative Ricci curvature. Then every positive function $u : M \to \mathbb{R}$ that is harmonic, $\Delta u = 0$, must be constant.*

1.3 Comparison Theorems

Laplacian Comparison Theorem. *Let $(\tilde{M}, \tilde{g})$ be an n-dimensional complete, simply connected Riemannian manifold of constant curvature K and (M, g) an n-dimensional complete Riemannian manifold with Ricci curvature bounded from below: $Ric \geq (n-1)K \cdot g$. Let $\tilde{p} \in \tilde{M}$ and $p \in M$ be fixed points, and denote by $\tilde{r}$ the geodesic distance function from $\tilde{p}$ to $\tilde{x}$ on $\tilde{M}$, and by r from from p to x on M; assume that the distance functions are differentiable in their arguments. If, for $x \in M$ and $\tilde{x} \in \tilde{M}$, we have $r(x) = \tilde{r}(\tilde{x})$ then*

$$\Delta r(x) \leq \tilde{\Delta} \tilde{r}(\tilde{x}),$$

where Δ and $\tilde{\Delta}$ denote the Laplace operators on (M, g) and $(\tilde{M}, \tilde{g})$, respectively. For a proof see the Appendix A.2.4 in [58].

From the Laplacian Comparison Theorem we have the following

Theorem. *Let (M, g) be an n-dimensional complete Riemannian manifold with Ricci curvature bounded from below by a constant $K \leq 0$. Then the geodesic distance function r satisfies*

$$r \Delta r(x) \leq (n-1)(1 + \sqrt{-K} \cdot r).$$

To state the following comparison Lemma about the *normal mapping*, we first recall two definitions from [37]. Up to the end of section 1.3, let Ω be an open subset of $\mathbb{R}^n$ with coordinates $(x^1, ..., x^n)$, and let $u : \Omega \to \mathbb{R}$. If E is a set, then $\mathcal{P}(E)$ denotes the class of all subsets of E.

The normal mapping. ([37], p.1). *The normal mapping of u, or subdifferential of u, is the set valued function $\partial u : \Omega \to \mathcal{P}(\mathbb{R}^n)$ defined by*

$$\partial u(x_0) = \{p \mid u(x) \geq u(x_0) + p \cdot (x - x_0), \quad for \quad all \quad x \in \Omega\}.$$

Given $E \subset \Omega$, we define $\partial u(E) := \bigcup_{x \in E} \partial u(x)$.

Viscosity solution. ([37], p.8). *Let $\Omega \in \mathbb{R}^n$ be a bounded domain with coordinates $(x^1, ..., x^n)$, let $u \in C(\Omega)$ be a convex function, and $f \in C(\Omega)$, $f \geq 0$. The function u is a viscosity subsolution (supersolution) of the equation* $\det\left(\frac{\partial^2 u}{\partial x^i \partial x^j}\right) = f$ *in Ω if, whenever a convex function $\phi \in C^2(\Omega)$ and $x_0 \in \Omega$ are such that*

$$(u - \phi)(x) \leq (\geq)(u - \phi)(x_0)$$

for all x in a neighborhood of x_0, then we must have

$$\det\left(\frac{\partial^2 \phi}{\partial x^i \partial x^j}\right)(x_0) \geq (\leq) f(x_0).$$

Normal mapping comparison Lemma. ([37], p.10). *Let $\Omega \subset \mathbb{R}^n$ be a bounded open set, and $u,\ v \in C(\bar{\Omega})$. If $u = v$ on $\partial\Omega$ and $v \geq u$ in Ω, then the normal mappings satisfy*

$$\partial v(\Omega) \subset \partial u(\Omega).$$

A comparison principle for Monge-Ampère equations. ([16] or [37], p.25). *Let Ω be a bounded open subset of $\mathbb{R}^n$, $n \geq 2$, and let $f \in C^0(\Omega)$ be a positive function. Assume that $w \in C^0(\bar{\Omega})$ is a locally convex viscosity subsolution (supersolution) of*

$$\det\left(\frac{\partial^2 w}{\partial x^i \partial x^j}\right) = f \quad in \quad \Omega,$$

and $v \in C^0(\bar{\Omega}) \cap C^2(\Omega)$ is a locally convex supersolution (subsolution) of

$$\det\left(\frac{\partial^2 v}{\partial x^i \partial x^j}\right) = f \quad in \quad \Omega.$$

Assume also that

$$w \leq v \quad (w \geq v) \quad on \quad \partial\Omega.$$

Then

$$w \leq v \quad (w \geq v) \quad on \quad \Omega.$$

1.4 The Legendre Transformation

Consider a locally strongly convex hypersurface $x : \Omega \to \mathbb{R}^{n+1}$, defined on a domain $\Omega \subset \mathbb{R}^n$ and given as graph of a strictly convex function

$$f : \Omega \to \mathbb{R}, \qquad x = (x^1, ..., x^n) \mapsto f(x) = f(x^1, ..., x^n).$$

Consider the *Legendre transformation* of f

$$\xi_i := \frac{\partial f}{\partial x^i}, \quad i = 1, 2, \ldots, n, \qquad u(\xi) := u(\xi_1, \ldots, \xi_n) := \sum x^i \frac{\partial f}{\partial x^i} - f(x)$$

and denote by Ω^* the *Legendre transform domain* of f, where $u : \Omega^* \to \mathbb{R}$ and

$$\Omega^* := \{ (\xi_1(x), ..., \xi_n(x)) \mid x \in \Omega \}.$$

Vice versa we have

$$x^i = \frac{\partial u}{\partial \xi_i}, \quad \text{and} \quad f(x) := \sum \xi_i \frac{\partial u}{\partial \xi_i} - u(\xi).$$

In the following we keep in mind the bijective relation $x \leftrightarrow \xi$ between corresponding points of the transformation and consider the functions $f = f(x)$ and $u = u(\xi)$ at such corresponding points, resp. This gives an involution of the relations. One calculates

$$\frac{\partial^2 u}{\partial \xi_i\, \partial \xi_j} = \frac{\partial x^i}{\partial \xi_j}, \quad \text{and} \quad \frac{\partial^2 f}{\partial x^i\, \partial x^j} = \frac{\partial \xi_i}{\partial x^j}.$$

It follows that the matrix

$$\left(\frac{\partial^2 u}{\partial \xi_i\, \partial \xi_j} \right)_\xi$$

is inverse to the matrix

$$\left(\frac{\partial^2 f}{\partial x^i\, \partial x^j} \right)_x.$$

We will use this transformation for the representation of graph hypersurfaces and the solution of Monge-Ampère equations. The fact that both matrices are inverse has advantages for calculations. We define two auxiliary functions ρ and Φ as follows

$$\left[\det\left(\frac{\partial^2 f}{\partial x^i \partial x^j}\right)\right]^{-\frac{1}{n+2}} =: \rho(x) = \rho(\xi) = \left[\det\left(\frac{\partial^2 u}{\partial \xi_i \partial \xi_j}\right)\right]^{\frac{1}{n+2}},$$

$$\sum f^{ij} \frac{\partial \ln \rho}{\partial x^i} \frac{\partial \ln \rho}{\partial x^j} =: \Phi(x) = \Phi(\xi) = \sum u^{ij} \frac{\partial \ln \rho}{\partial \xi_i} \frac{\partial \ln \rho}{\partial \xi_j}.$$

As above f_{ij} denotes the components of the Hessian matrix and $f^{ij} f_{jk} = \delta^i_k$ gives its inverse matrix. The two expressions for ρ show that we can consider ρ as a function in terms of the x-coordinates and also as a function in terms of the ξ-coordinates; analogously, this view point holds for Φ, too.

Chapter 2

Local Equiaffine Hypersurfaces

2.1 Hypersurfaces in Unimodular Affine Space

In this chapter we summarize the local equiaffine hypersurface theory. We use Cartan's calculus of exterior forms and a standard local calculus. As the monograph [58] contains a detailed development of the theory in Cartan's calculus, in a local notation and also in Koszul's calculus, our introduction here has a more condensed form.

A reader who is familiar with affine hypersurface theory in Cartan's calculus can skip chapter 2. For a reader not familiar with that, chapter 2 offers a guided survey; we give some proofs that can be used as introductory exercises, while more details can be found in the monographs [58], [73], [88].

2.1.1 *The ambient space*

In order to define the *unimodular affine structure* of a space one uses the associated vector space; that means:
Let A^{n+1} denote the real affine space of dimension $n+1$ and V the associated real vector space of the same dimension, V^* its dual space. They are equipped with the following structures:

- $\langle\, ,\, \rangle : V^* \times V \to \mathbb{R}$ the canonical scalar product;
- there is a one-dimensional vector space of determinant forms over V; determinant forms are denoted by Det, we can use them as volume forms; correspondingly there is a one-dimensional vector space of dual determinant forms over V^*, they are denoted by Det^*;
- we denote the directional derivation in V and V^* by the same symbol $\bar{\nabla}$.

The three structures satisfy the standard compatibility conditions. Thus the structure of an affine space is defined using its associated vector space. Considering A^{n+1} as a differentiable manifold,there is a tangent space T_pA^{n+1} at each point $p \in A^{n+1}$. The duality allows to extend the concept of the well known *cross product construction* in the Euclidean 3-space to the affine setting in any dimension. Consider a linearly independent family $\{w_1, ..., w_n\} \subset V$ and $W := \text{span}\{w_1, ..., w_n\}$; via duality there is a 1-dimensional subspace $W^* \subset V^*$ such that $w^*(w_i) = 0$ for any $w^* \in W^*$. A basis for W^* can be explicitly calculated from the cross product construction as follows: For a fixed non-trivial determinant form Det, we define the *cross product* $[w_1, ..., w_n]$

$$[\,, ..., \,] : \prod_n V \to V^*$$

by

$$\langle [w_1, ..., w_n], z\rangle := Det(w_1, ..., w_n, z) \qquad \forall z.$$

Affine mappings that preserve the affine structure are defined via linear mappings between the associated vector spaces, and additional translations in A^{n+1}. An *affine transformation* $\mathfrak{A} : A^{n+1} \to A^{n+1}$ is an affine mapping of maximal rank, that means the associated linear mapping $A : V \to V$ is an automorphism, $A \in GL(n+1, \mathbb{R})$. Fixing a *coordinate system*, that is an *origin* in A^{n+1} and a *basis* of V, we can express an affine transformation in matrix notation:

$$\tilde{x} = A \cdot x + d,$$

where $\tilde{x}, x \in \mathbb{R}^{n+1}$ denote the coordinate vectors of points in A^{n+1}, and $d \in \mathbb{R}^{n+1}$ describes a translation. *Affine geometry* studies geometric properties of subsets of A^{n+1} that are invariant under affine transformations.

The unimodular space. In case we fix a determinant form Det as volume form over V, its dual determinant form is denoted by Det^*. An affine transformation with

$$\det A = 1, \quad \text{i.e.,} \quad A \in SL(n+1, \mathbb{R}),$$

is called *unimodular* or *equiaffine*; here $\det A$ denotes the determinant of the matrix A. *Equiaffine geometry* is the study of geometric properties that are invariant under unimodular transformations; their invariants are called *unimodular* or *equiaffine*. *Volume* is a unimodular invariant, while distance and angle are *not* preserved under unimodular transformations; they are preserved under Euclidean motions. We will use the notation A^{n+1} also for the *unimodular affine space* as the context will be clear. In particular, in all sections of Chapter 2, A^{n+1} denotes the *unimodular affine space*.

Notational convention. Our convention for the range of indices is as follows:

$$1 \leq \alpha, \beta, \gamma, \cdots \leq n+1,$$

$$1 \leq i, j, k, \cdots \leq n;$$

as usual we adopt Einstein's summation convention.

2.1.2 *Affine hypersurfaces*

A hypersurface consists of an n-dimensional differentiable manifold M and an immersion

$$x : M \to A^{n+1}. \tag{2.1.1}$$

In a short terminology we write x for a hypersurface. We fix a coordinate system in A^{n+1}; then it is standard to denote the position vector of the hypersurface x with respect to a fixed origin with the same symbol x.
A *unimodular affine frame*, or simply a *frame*, is a point $p \in A^{n+1}$ together with $n+1$ tangent vectors $e_1, \cdots, e_{n+1} \in T_pA^{n+1}$ satisfying the condition

$$Det(e_1, \cdots, e_{n+1}) = 1. \tag{2.1.2}$$

The importance of frames in affine geometry lies in the fact that there is exactly one unimodular affine transformation carrying one frame into another.
In the space of all unimodular frames we consider the expressions

$$dx = \sum_\alpha \omega^\alpha e_\alpha, \tag{2.1.3}$$

$$de_\alpha = \sum_\beta \omega_\alpha^\beta e_\beta. \tag{2.1.4}$$

The coefficients are differentiable, and the tangent fields e_α to A^{n+1} define a *frame field.* The coefficients ω^α, ω_α^β are called the *Maurer-Cartan forms* of $SL(n+1, \mathbb{R})$. Differentiating (2.1.2) and using (2.1.4) we get

$$\sum_\alpha \omega_\alpha^\alpha = 0. \tag{2.1.5}$$

Exterior differentiation of (2.1.3) and (2.1.4) gives the *structure equations* of A^{n+1} or the *Maurer-Cartan equations* of $SL(n+1, \mathbb{R})$:

$$d\omega^\alpha = \sum_\beta \omega^\beta \wedge \omega_\beta^\alpha, \tag{2.1.6}$$

$$d\omega_\alpha^\beta = \sum_\gamma \omega_\alpha^\gamma \wedge \omega_\gamma^\beta. \tag{2.1.7}$$

Now the important step is that we restrict to the submanifold of frames such that x lies on the hypersurface, and $e_1, \cdots, e_n$ span the tangent hyperplane at x. Then

$$\omega^{n+1} = 0 \tag{2.1.8}$$

and the equation (2.1.6) gives

$$\sum_i \omega^i \wedge \omega_i^{n+1} = 0. \tag{2.1.9}$$

2.2 Structure Equations and Berwald-Blaschke Metric

Let $x : M \to A^{n+1}$ be a C^∞ hypersurface. Since the investigation is local, we may identify M with $x(M)$. Then the tangent space T_pM at $p \in M$ can be identified with an n-dimensional subspace, denoted by $T_{x(p)}M$, of the tangent space $V = T_{x(p)}A^{n+1}$. We can choose a local unimodular affine frame field $\{p; e_1, \cdots, e_{n+1}\}$ on M such that $p \in M$ and $e_1, \cdots, e_n \in T_pM$. We call such a frame *adapted to M at p* (shortly an *adapted frame*); with respect to such a frame we have (2.1.3-4); these equations are called *moving frame equations* for the hypersurface. Again we consider the restrictions to M and T_pM as in (2.1.8-9), and the forms and fields to be locally differentiable on M. We apply Cartan's lemma to (2.1.9) and get

$$\omega_i^{n+1} = \sum h_{ij}\omega^j,$$

where the local coefficients are symmetric and locally differentiable on M:

$$h_{ij} = h_{ji}.$$

2.2.1 *Structure equations - preliminary version*

For an adapted frame field we arrive at a preliminary version of the structure equations associated to the moving frame equations:

Gauß equation $$de_i = \sum_j \widetilde{\omega}_i^j\, e_j + \sum h_{ij}\omega^j e_{n+1},$$

Weingarten equation $$de_{n+1} = \sum \omega_{n+1}^i\, e_i + \omega_{n+1}^{n+1}\, e_{n+1}.$$

In the following we will discuss the coefficients of the structure equations. First we consider the quadratic differential form

$$\sum \omega^i\, \omega_i^{n+1} = \sum h_{ij}\, \omega^i \omega^j. \qquad (2.2.1)$$

We assume M to be oriented and state the following Lemma; for the proof we refer to our detailed exposition in [58].

Lemma.

(i) *The expression (2.2.1) is invariant under unimodular affine transformations in A^{n+1}, although the expression depends on the choice of the local frame field.*

(ii) *The rank of the quadratic differential form (2.2.1) is an affine invariant.*

(iii) *Assume that $rank(h_{ij}) = n$, thus $H := det(h_{ij}) \neq 0$. Define*

$$G_{ij} := |H|^{\frac{-1}{n+2}}\, h_{ij}, \qquad (2.2.1.a)$$

and

$$G := \sum G_{ij}\omega^i\omega^j. \qquad (2.2.1.b)$$

Then G is independent of the choice of the local unimodular affine frame field, moreover it is an equiaffinely invariant form on the hypersurface.

A hypersurface with $rank\, G = n$ is called *non-degenerate*. The equiaffinely invariant form G can be used as semi-Riemannian metric. G is definite if and only if the hypersurface is locally strongly convex. Then, by an appropriate choice of the orientation, we can assume G to be positive definite, i.e., G is a Riemannian metric on M. Nowadays, in a standard terminology, one simply calls G the *Blaschke metric* of the hypersurface, in a short terminology one speaks about a *Blaschke hypersurface*.

Notational convention. Using G as semi-Riemannian metric, its Levi-Civita connection induces a covariant differentiation. With respect to G and $\nabla := \nabla(G)$ we use the local standard notation that we introduced in the foregoing chapter.

(i) In a coordinate notation, the matrix associated to G is written (G_{ij}) and its inverse matrix by (G^{ik}): the operations of lowering and raising indices are now defined with respect to the metric tensor G; as already stated, in the local notation we adopt the Einstein summation convention;

(ii) we use orthonormal frames $\{e_1, ..., e_n\}$ and their dual coframes $\{\omega^1, ..., \omega^n\}$;

(iii) the Riemannian volume form of the Blaschke metric G is given by

$$dV = |H|^{\frac{1}{n+2}}\ \omega^1 \wedge ... \wedge \omega^n;$$

(iv) the *connection forms* $\widetilde{\omega}_i^j$ in the foregoing version of the Gauß structure equations define a connection $\widetilde{\nabla}$ on M:

$$\widetilde{\omega}_i^k = \sum \widetilde{\Gamma}_{ij}^k \omega^j,$$

it is called the *induced connection*; in the local notation one can express this connection in terms of its coefficients, the *Christoffel symbols* $\widetilde{\Gamma}_{ij}^k$; $\widetilde{\nabla}$ is a torsion free, Ricci-symmetric connection; its Christoffel symbols are symmetric in (i, j);

(v) we denote the Levi-Civita connection of G by ∇ and its Christoffel symbols by Γ_{ij}^k; we have

$$\omega_i^k = \sum \Gamma_{ij}^k \omega^j;$$

(vi) as both connections are torsion free, we get the symmetry (see also [58], p.44):

$$\widetilde{\Gamma}_{ij}^k - \Gamma_{ij}^k = \widetilde{\Gamma}_{ji}^k - \Gamma_{ji}^k.$$

This gives:

Lemma. *The local coefficients*

$$A_{ij}^k := \widetilde{\Gamma}_{ij}^k - \Gamma_{ij}^k \tag{2.2.2}$$

define a symmetric (1,2)-tensor field A.

2.2.2 *Covariant Gauß equations - preliminary*

We use the moving frame equation for x and its Gauß structure equation above to rewrite the Gauß structure equation in a local covariant notation (see section 1.1.1) - still in a preliminary version:

$$x_{,ij} = \sum_k A_{ij}^k \, e_k + h_{ij} \, e_{n+1}. \tag{2.2.3}$$

As the metric is a unimodular invariant, the left hand side is invariant under unimodular transformations, it only depends on the choice of the adapted frame; we aim at unimodularly invariant terms on the right hand side; it is obvious that we will insert the Blaschke metric into the second term (see the next section).

2.3 The Affine Normalization

We recall the normalization for a hypersurface x in Euclidean space and some of its basic properties. We identify V and V^* and write the inner product by $\langle\, , \rangle : V \times V \to \mathbb{R}$. We consider a fixed point $x(p) \in x(M)$ where $p \in M$.

(i) The Euclidean unit normal $\mu(p)$ at $x(p)$ extends a basis of the tangent hyperplane $T_{x(p)}M$ to a basis of $T_{x(p)}A^{n+1}$.
(ii) $\mu(p)$ determines $T_{x(p)}M$.
(iii) We have $\langle \mu, \mu \rangle = 1$ and thus $\langle \mu, d\mu(v) \rangle = 0$ for any $v \in T_{x(p)}M$; this implies that $d\mu(v)$ is tangential for any $v \in T_{x(p)}(M)$.
(iv) The deviation $d\mu$ induces the *Euclidean shape operator* and thus gives rise to the understanding of *Euclidean extrinsic curvature.*
(v) In case that the Euclidean shape (Weingarten) operator has maximal rank, μ defines the *Euclidean Gauß map*, an immersion $\mu : M \to S^n$.
(vi) The pair (x, μ) is invariant under Euclidean motions.

Analogously we aim to find an affine invariant normalization. Every transversal field to an affine hypersurface extends a basis of the tangent hyperplane $T_{x(p)}M$; but no transversal field fixes the tangent hyperplane as long as the ambient space has no Euclidean structure. Instead, to fix the tangent hyperplane at a point, we consider the dual vector space V^* of $V := T_{x(p)}A^{n+1}$. By duality there is a 1-dimensional subspace $C_{x(p)} \subset V^*$ at any $p \in M$, defining a line bundle along M. This bundle is called the *conormal line bundle.*

2.3.1 *The affine normal*

In a first step we aim to find a field transversal to $x(M)$ that is invariant under unimodular transformations.

Consider a non-degenerate hypersurface $x : M \to A^{n+1}$ and fix a coordinate system in A^{n+1}. According to section 2.1.2 the position vector $x = (x^1, \cdots, x^{n+1})$ with component functions x^i is a vector valued function. We define a vector field $Y : M \to V$ by

$$Y := \tfrac{1}{n}\Delta x = \tfrac{1}{n}(\Delta x^1, \cdots, \Delta x^{n+1}), \tag{2.3.1}$$

where Δ denotes the Laplacian with respect to the Blaschke metric. As the metric G is an equiaffine invariant, Y is equiaffinely invariant by construction.
We choose a local adapted frame field $\{x; e_1, \cdots, e_{n+1}\}$ on M and compute Δx:

$$\begin{aligned} Y &= \tfrac{1}{n}\Delta x = \tfrac{1}{n}\sum G^{ij} x_{,ij} \\ &= \tfrac{1}{n}\sum G^{ij}\left(A_{ij}^k\, e_k + h_{ij} e_{n+1}\right) \\ &= \tfrac{1}{n}\sum G^{ij} A_{ij}^k\, e_k + |H|^{\frac{1}{n+2}} e_{n+1}. \end{aligned} \tag{2.3.2}$$

Hence

$$Det\,(e_1, \cdots, e_n, Y) = |H|^{\frac{1}{n+2}} \neq 0. \tag{2.3.3}$$

The last two equations are the basis for a proof of the following results for the field Y defined in (2.3.1).

Apolarity Condition. *The following three properties are equivalent:*

(a) Y *is parallel to* e_{n+1}*;*

(b) $G^{ij} A_{ij}^k = 0 \quad for \quad k = 1, ..., n;$

(c) $\omega_{n+1}^{n+1} + \frac{1}{n+2}\, d \ln |H| = 0.$

The vector Y*, satisfying one of the conditions (a)-(c), is called the (equi-)affine normal of* x*.*
For the proof we refer to [58], section 1.2.

Remarks. (i) When $x : M \to A^{n+1}$ is locally strongly convex, from the above calculation one can easily see that Y always points to the concave side of $x(M)$.
(ii) The geometric meaning of the apolarity condition is the following: Both, the Levi-Civita and the induced connection, have symmetric Ricci tensors. Thus both connections ∇ and $\widetilde{\nabla}$ admit parallel volume forms; in case of the Levi-Civita connection it is the Riemannian volume form. Now the apolarity condition, written in the form

$$G^{ij}\Gamma_{ij}^k = G^{ij}\widetilde{\Gamma}_{ij}^k,$$

also implies that both volume forms coincide (modulo a non-zero constant factor). This geometric argument was chosen by H. Flanders and K. Nomizu to introduce Y as *affine normal*; [32], [72].
(iii) While the pair (x, Y) with Y as affine normal field is equiaffinely invariant, the lines generated by the affine normals define a line bundle; this line bundle is *affinely invariant.* This line bundle is called the *affine normal bundle.*

Remark. When e_{n+1} is parallel to Y, the last formula (c) and

$$de_{n+1} = \sum \omega_{n+1}^{i}\, e_i \;+\; \omega_{n+1}^{n+1}\, e_{n+1}$$

give

$$dY = |H|^{\frac{1}{n+2}} \sum \omega_{n+1}^{i}\, e_i. \tag{2.3.4}$$

We will frequently need condition (c) in the apolarity condition above for explicit calculations.

Equiaffine frames. From now on we shall choose an adapted frame field $\{x; e_1, \cdots, e_n, e_{n+1}\}$ such that e_{n+1} is parallel to Y. We call such a frame an *equiaffine frame*; so an equiaffine frame has the three properties:

(i) it is unimodular,

(ii) $e_1, \cdots, e_n$ are tangential,

(iii) e_{n+1} is parallel to the affine normal vector Y.

This choice implies the apolarity condition and

$$Y = |H|^{\frac{1}{n+2}}\, e_{n+1}. \tag{2.3.5}$$

Moreover, (2.3.4) states the *Weingarten equation* for Y.

2.3.2 *Affine shape operator and affine extrinsic curvature*

For a hypersurface in Euclidean space the Weingarten equation for the unit normal implicitly defines the Euclidean *shape* or *Weingarten operator*; from this we get the extrinsic curvature functions.

For an arbitrary $p \in M$, equation (2.3.4) states that $dY(v)$ is tangential to $x(M)$ for any $v \in T_pM$. This situation suggests to search for an affine analogue of the Euclidean Weingarten operator.

Let $x; e_1, \cdots, e_n, e_{n+1}$ be an equiaffine frame on M. Exterior differentiation of (2.3.4) gives

$$\sum \omega_{n+1}^{i} \wedge \omega_{i}^{n+1} = 0. \tag{2.3.6}$$

Since M is non-degenerate, the forms $\omega_1^{n+1}, \omega_2^{n+1}, \cdots, \omega_n^{n+1}$ are linearly independent. Cartan's lemma and (2.3.6) imply

$$\omega_{n+1}^{i} = -\sum l^{ij} \omega_j^{n+1} \tag{2.3.7}$$

where the coefficients, implicitly defined in (2.3.7), are symmetric:

$$l^{ij} = l^{ji}.$$

We insert the relation $\omega_i^{n+1} = \sum h_{ij}\,\omega^j$ into (2.3.7) and obtain

$$\omega_{n+1}^i = -\sum_j l_j^i\,\omega^j, \tag{2.3.8}$$

$$l_j^i = \sum_k h_{jk}\,l^{ki}.$$

The associated quadratic differential form reads

$$B = -\sum_i \omega_{n+1}^i\,\omega_i^{n+1} = \sum_{i,j} l^{ij}\,\omega_i^{n+1}\,\omega_j^{n+1}$$
$$= \sum l^{ij}\,h_{ik}\,h_{jl}\,\omega^k\,\omega^l := \sum B_{ij}\,\omega^i\,\omega^j. \tag{2.3.9}$$

Exercise. **(i)** The quadratic differential form B is invariant under a change of frames keeping the affine normal field fixed.
(ii) The quadratic differential form B is symmetric.

The Weingarten form. We call the symmetric quadratic form B *(equi)-affine Weingarten form.* The symmetry of B implies that the associated operator $B^\sharp$, implicitly defined by $G(B^\sharp v, w) := B(v,w)$, is self adjoint with respect to the Blaschke metric G; it is called the *(equi)-affine shape* or *Weingarten operator.* On locally strongly convex hypersurfaces, where G is (positive) definite, the eigenvalues $\lambda_1, \lambda_2, \cdots, \lambda_n$ of $B^\sharp$ are real; they are unimodular invariants and are called *(equi)-affine principal curvatures.* The associated eigendirections are called *(equi)-affine principal curvature directions.* In a local notation, we write the coefficients of $B^\sharp$ also by $|H|^{\frac{1}{n+2}} l_i^j =: B_i^j$.

The affine extrinsic curvature functions. On a non-degenerate hypersurface consider the characteristic polynomial of $B^\sharp$; its coefficients are the (non-normed) *affine extrinsic curvature functions.* On a locally strongly convex hypersurface they coincide with the elementary symmetric functions of the eigenvalues:

$$\binom{n}{r} L_r := \sum_{1\le i_1<\cdots<i_r\le n} \lambda_{i_1}\cdots\lambda_{i_r}, \qquad r = 1,2,\cdots,n, \qquad L_0 := 1. \tag{2.3.10}$$

We call L_1 the *(equi)-affine mean curvature* and L_n the *(equi)-affine Gauß-Kronecker curvature.*

Theorem. *Let x be non-degenerate and let its dimension n be even. Then one can calculate the affine Gauß-Kronecker curvature from the induced connection $\widetilde{\nabla}$.*

For a proof see [74].

2.3.3 *The affine conormal*

In the introduction to section 2.3 we listed elementary properties of the Euclidean normalization of a hypersurface. In section 2.3.1 we defined the affine normal Y. The pair (x, Y) is invariant under unimodular transformations of A^{n+1}. In analogy to the Euclidean unit normal the transversal field Y has the property that $dY(v)$ is tangential to $x(M)$ for any tangent vector $v \in M$. But Y does *not* fix the tangent plane. We recall the notion of the *conormal line bundle* along M and call any nowhere vanishing section of this bundle a *conormal field* on M. We are going to search for a conormal field that is invariant under unimodular transformations.

First let us recall some elementary facts from multilinear algebra.
Let $\eta_1, \eta_2, \cdots, \eta_{n+1}$ be a basis of V. Then there exists a canonical isomorphism from the vector space of exterior n-forms to V^*, which is given by

$$i : \bigwedge^n V \to V^* \quad \text{where} \quad i(\alpha)(v) = a, \quad \alpha \in \bigwedge^n V, \quad a \in \mathbb{R}$$

if and only if

$$\alpha \wedge v = a \cdot \eta_1 \wedge \eta_2 \wedge \cdots \wedge \eta_{n+1}.$$

Thus we can identify $\bigwedge^n V$ and V^* via this isomorphism.

Definition. Let $x : M \to A^{n+1}$ be a non-degenerate hypersurface. For every $p \in M$ and for $V = T_{x(p)}A^{n+1}$, there exists a unique $U \in V^*$ satisfying the following two conditions:

$$\langle U, dx(v)\rangle = 0, \quad v \in T_pM, \qquad \langle U, Y\rangle = 1.$$

The vector $U \in V^*$ is called the *affine conormal vector* of M at p.
As above identify V and $T_{x(p)}A^{n+1}$ and consider an equiaffine frame $\{x; e_1, e_2, \cdots, e_{n+1}\}$. Then U can be identified with

$$|H|^{\frac{-1}{n+2}}\, e_1 \wedge e_2 \wedge \cdots \wedge e_n. \tag{2.3.11}$$

This expression corresponds to the cross product construction via duality, stated in section 2.1.1.

Covariant structure equations for the conormal. *The conormal of a non-degenerate hypersurface x satisfies the system of vector valued PDEs:*

$$U_{,ij} = -\sum A^k_{ij}\, U_k - B_{ij}\, U \tag{2.3.12}$$

and the Schrödinger type PDE

$$\Delta U = -\, nL_1\, U. \tag{2.3.13}$$

For a proof see section 1.3.1 in [58].

Lemma. (a) *On a non-degenerate hypersurface we have*

$$\langle U_i, e_j\rangle = -\, G_{ij}.$$

In particular, this implies rank $dU = n$.
Moreover, in a short notation, we have the following linear systems of equations:
(b) *For U given at a point, the system*

$$\langle U, Y\rangle = 1, \qquad \langle dU, Y\rangle = 0$$

uniquely determines Y.
(c) *Vice versa, for Y given, the system*

$$\langle U,Y\rangle = 1, \qquad \langle U,dY\rangle = 0$$

uniquely determines U.
(d) *As a consequence, at any $p \in M$, the relation $Y \leftrightarrow U$ is bijective.*

Proof. U fixes the tangent plane, thus, for any tangential frame, we have $\langle U,e_i\rangle = 0$. Exterior differentiation of the equation $\langle U,Y\rangle = 1$ gives

$$0 = \sum \langle U_i, Y\rangle\, \omega^i + |H|^{\frac{1}{n+2}} \sum \langle U, e_i\rangle\, \omega^i_{n+1} = \sum \langle U_i, Y\rangle\, \omega^i.$$

Hence

$$\langle U_i, Y\rangle = 0, \qquad i = 1, 2, \cdots, n.$$

Analogously an exterior differentiation of the equation $\langle U, e_i\rangle = 0$ implies

$$\begin{aligned} 0 &= \sum \langle U_j, e_i\rangle\, \omega^j + \sum \langle U\,,\, \omega^j_i e_j + \omega^{n+1}_i e_{n+1}\rangle \\ &= \sum \left(\langle U_j,\, e_i\rangle + G_{ij}\right) \omega^j. \end{aligned}$$

This gives the assertion. ∎

2.3.4 *The conormal connection*

In (2.2.2) we defined the difference tensor

$$A := \widetilde{\nabla} - \nabla.$$

One easily verifies that also $\nabla^* := \nabla - A$ defines another torsion free, Ricci-symmetric connection. Using this connection one can rewrite the Gauß conormal structure equations from (2.3.12) in the local form

$$U_{ij} = \sum \Gamma^{*\,k}_{ij}\, U_k \;-\; B_{ij}\, U$$

with Christoffel symbols Γ^{*k}_{ij}.
Exercise: The connection ∇^* is projectively flat; see [73], p.17.

2.3.5 *Affine Gauß mappings*

A *unimodular* or *Blaschke* hypersurface is a triple (x, U, Y) with (U, Y) as *equiaffine normalization* of x. Recall the statements about the properties of the Euclidean normalization in the beginning of section 2.3. The subsections following these statements show that the affine normalization allows to list properties similar to the Euclidean case.
For a non-degenerate hypersurface we know that the mapping $U : M \to V^*$ always has maximal rank, thus it defines an immersion; moreover, it is easy to show that its position vector, again denoted by U, is always transversal to $U(M)$. This immersion itself is non-degenerate if and only if the Weingarten form B has maximal

rank. In this case we call the mapping $U : M \to V^*$ the *affine conormal Gauß map*, and we can use B as unimodular metric of this hypersurface; then we call $U(M)$ *affine conormal indicatrix.*
The situation for $Y : M \to V$ is different: We have $rank\, Y = n$ if and only if $rank\, B = n$, and only in this case Y is an immersion. But then Y is also transversal to $Y(M)$ and this hypersurface is also non-degenerate; again we can use B as unimodular metric. We call $Y : M \to V$ the *affine normal Gauß map* and the hypersurface $Y(M)$ the *affine normal indicatrix.* The fact that both *affine indicatrices* have the same unimodular metric in particular implies that the *affine conormal indicatrix* is locally strongly convex if and only if the *affine normal indicatrix* is locally strongly convex.

Affine Gauß maps and Euclidean structure. In the case $rank\, B = n$ it is often convenient to consider the two hypersurfaces, defined from the affine Gauß maps, as follows: We consider a Euclidean inner product $\langle\ ,\ \rangle : V \times V \to \mathbb{R}$ on V and identify V and V^* as usual. The three relations

$$\langle U, Y\rangle = 1, \quad \langle U, dY\rangle = 0, \quad \langle dU, Y\rangle = 0$$

imply that both affine Gauß indicatrices are a *polar pair*, that means they correspond via an *inversion at the unit sphere.* For an equiaffine frame $\{e_1, ..., e_n\}$ we can calculate the conormal with the cross product construction:

$$U = [\det(G)]^{-\frac{1}{2}} \cdot [e_1, ..., e_n].$$

For the affine normal indicatrix we have the relation (see [58], p.52, (1.2.3.10))

$$[Y_1, ..., Y_n] = (-1)^n \det B \cdot [e_1, ..., e_n].$$

Using the Euclidean structure of V, we can express the conormal in terms of the Euclidean unit normal μ of x:

$$U = |K|^{\frac{1}{n+2}} \cdot \mu.$$

Here K is the *Euclidean Gauß-Kronecker curvature* of the hypersurface x; see [88], section 6.2.4.

2.4 The Fubini-Pick Form

In the covariant form of the structure equations there appears the symmetric (1,2)-tensor field A which was defined in (2.2.2). As before we use an equiaffine frame $\{x; e_1, \cdots, e_n, e_{n+1}\}$ on M. To A associated there is the *cubic form* or *Fubini-Pick form*:

$$A^\flat := \sum A_{ijk}\, \omega^i \omega^j \omega^k \tag{2.4.1}$$

with local components

$$A_{ijk} := \sum G_{il}\, A^l_{jk}.$$

As usual, in a local notation, we simply write $A^\flat_{ijk} =: A_{ijk}$. In case the meaning is clear one also sometimes simplifies the notation for $A^\flat$, just writing the cubic form by A. We will prove

2.4.1 *Properties of the Fubini-Pick form*

Lemma. *We have:* **(i)**

$$A_{ijk} = -\tfrac{1}{2}|H|^{\frac{-1}{n+2}}\, h_{ijk}, \tag{2.4.2}$$

where $\qquad \sum h_{ijk}\,\omega^k := dh_{ij} + h_{ij}\,\omega^{n+1}_{n+1} - \sum h_{ik}\,\omega^k_j - \sum h_{kj}\,\omega^k_i.$

(ii) $A^\flat$ *is totally symmetric:*

$$A_{ijk} = A_{jik} = A_{ikj}. \tag{2.4.3}$$

(iii) *The cubic form* $A^\flat$ *is invariant under unimodular transformations.*

(iv) *By definition the difference tensor* A *measures the deviation of the two connections* ∇ *and* $\widetilde{\nabla}$.

For a proof see [58], section 1.2.2.

2.4.2 *The Pick invariant*

We recall the Gauß structure equations for x from section 2.2.2; in the covariant form below there appear A and G as coefficients, both are equiaffinely invariant tensor fields. The simplest scalar invariant of the metric and the cubic form is defined by

$$J := \tfrac{1}{n(n-1)} \sum G^{il} G^{jm} G^{kr} A_{ijk}\, A_{lmr} = \tfrac{1}{n(n-1)} \|A\|^2,$$

where the tensor norm $\|\cdot\|$ is taken with respect to the Blaschke metric G. J is called the *Pick invariant.* If $A = 0$ then trivially $J = 0$; on locally strongly convex hypersurfaces G is (positive) definite, we then have the implication: $J = 0 \Rightarrow A = 0$.

2.4.3 *Structure equations - covariant notation*

We recall the preliminary versions of the structure equations in sections 2.2.1 and 2.2.2. We clarified that the Blaschke metric G, the cubic form $A^\flat$, the affine normal Y and the affine shape operator are invariant under unimodular transformations. Thus we rewrite the structure equations in terms of G-covariant differentiation and with equiaffinely invariant coefficients as follows:

Gauß equation for x $\qquad x_{,ij} = \sum A^k_{ij}\, e_k + G_{ij}\, Y.$

Weingarten equation $\qquad Y_j = -\sum B^i_j\, e_i.$

Gauß equation for U $\qquad U_{,ij} = -\sum A^k_{ij}\, U_k - B_{ij}\, U.$

2.4.4 *The affine support function*

Definition. Let b be a fixed vector in V. The function $\Lambda : M \to \mathbb{R}$ defined by

$$\Lambda(p) := \langle U, b - x(p)\rangle, \quad p \in M,$$

is called the *affine support function* relative to the vector $b \in \mathbb{R}^{n+1}$.
The support function satisfies PDEs that play an important role for global investigations. We are going to compute the Laplacian of Λ. Let $d\Lambda = \sum \Lambda_i \omega^i$. From the definition of the conormal it follows that

$$d\Lambda = \langle dU, b - x\rangle - \langle U, dx\rangle = \sum \langle U_i, b - x\rangle \, \omega^i.$$

Hence

$$\Lambda_i = \langle U_i , b - x\rangle.$$

We calculate the second covariant derivative (called the *covariant Hessian* of $\Lambda(x)$)

$$\begin{aligned}\sum \Lambda_{,ij}\, \omega^j &= d\Lambda_i - \sum \omega_i^j \Lambda_j \\ &= \langle dU_i, b - x\rangle - \langle U_i, dx\rangle - \sum \omega_i^j \Lambda_j \\ &= \langle dU_i - \sum \omega_i^j U_j, b - x\rangle - \sum \langle U_i, e_j\rangle \, \omega^j \\ &= \sum \left(\langle U_{,ij}, b - x\rangle - \langle U_i, e_j\rangle\right) \omega^j.\end{aligned}$$

Therefore, the G-covariant Hessian of Λ satisfies

$$\Lambda_{,ij} = \langle U_{,ij}, b - x\rangle - \langle U_i, e_j\rangle. \tag{2.4.4}$$

The covariant conormal structure equations and the apolarity condition imply:

Covariant PDEs for the support function.

$$\Lambda_{,ij} = -\sum A_{ij}^k \Lambda_k - B_{ij}\Lambda + G_{ij}. \tag{2.4.5}$$

$$\Delta\Lambda + nL_1\Lambda = n. \tag{2.4.6}$$

Note that the equations have the same form for any $b \in V$. Moreover, (2.4.5) implies that on any hyperovaloid there exist points such that the Weingarten form is (positive) definite.

2.5 Integrability Conditions

2.5.1 *Integration via moving frames*

Like the structure equations in Euclidean hypersurface theory, the affine structure equations of Weingarten and Gauß for a hypersurface

$$\begin{aligned} de_{n+1} &= \sum \omega_{n+1}^i \, e_i, \\ de_i &= \sum \omega_i^j \, e_j + \omega_i^{n+1} \, e_{n+1} \end{aligned} \tag{2.5.1}$$

give a linear system of first order PDEs for a local frame $\{e_1, \cdots, e_n, e_{n+1}\}$. The coefficients define linear forms that are related to the two connections ∇, $\widetilde{\nabla}$, the quadratic form G, the cubic form A and the Weingarten operator:

$$\omega_{n+1}^i = -\sum l^{ik}\,\omega_k^{n+1} = -\sum l_k^i\,\omega^k, \qquad \omega_i^j = \sum \Gamma_{ik}^j\,\omega^k,$$

$$\omega_i^{n+1} = \sum h_{ij}\,\omega^j, \quad \text{where } G_{ij} = |H|^{\frac{-1}{n+2}}\,h_{ij}.$$

From the integration theory for such linear systems we know that there exists at most one solution $\{e_1, \cdots, e_n, e_{n+1}\}$ for given coefficients and given initial values. In particular, such a solution determines

$$dx = \sum \omega^i\,e_i\,,$$

and a second integration locally gives the hypersurface x itself. Thus, roughly speaking, the coefficients must contain all geometric information about the hypersurface x. The existence of a solution of the system depends on the fact that the coefficients satisfy integrability conditions. We are going to clarify this.
Choose a local equiaffine frame field $\{x; e_1, \cdots, e_n, e_{n+1}\}$ over M such that $e_{n+1} = Y$, $G_{ij} = \delta_{ij}$.

Theorem. *The integrability conditions of the system*

(a) $\quad dx = \sum \omega^i\,e_i,$

(b) $\quad de_i = \sum \omega_i^j\,e_j + \omega_i^{n+1}\,e_{n+1},$

(c) $\quad de_{n+1} = \sum \omega_{n+1}^i\,e_i,$

(d) $\quad \omega_i^{n+1} = \omega^i, \qquad \omega_{n+1}^{n+1} = 0$

read

(e) $\quad \sum \omega_i^i = 0,$

(f) $\quad d\omega^i = \sum \omega^j \wedge \omega_j^i,$

(g) $\quad d\omega_i^j = \sum \omega_i^k \wedge \omega_k^j + \omega_i^{n+1} \wedge \omega_{n+1}^j, \qquad \omega_i^{n+1} = \omega^i,$

(h) $\quad d\omega_{n+1}^i = \sum \omega_{n+1}^j \wedge \omega_j^i.$

The equations (e)-(h) between the linear differentiable forms ω^i, ω_i^j, ω_{n+1}^i *are sufficient for the integration of the systems (a)-(d).*

Proof. Since $\omega^{n+1} = 0$, the proof follows from the relations (a)-(d); apply the rules of exterior differentiation from section 1.1.1. ■

Terminology. In the *terminology of moving frames* the integrability conditions (e)-(h) are called *structure equations*, which means that they are necessary and sufficient for the existence of the hypersurface structure. But from our foregoing study we know that also the equations of Gauß and Weingarten are called *structure* *equations*, as their coefficients contain all information on the geometry of the hypersurface. To avoid any misunderstanding, we will use the terminology *integrability conditions* for the system (*e*)-(*h*) and *structure equations* for the equations of Gauß and Weingarten.

2.5.2 *Covariant form of the integrability conditions*

The integrability conditions give information about the dependence of the invariants that appear in the structure equations. We are going to express these conditions in terms of the quadratic and cubic forms G, B, and A. We state:

Integrability Conditions. *In covariant form the integrability conditions read:*

$$A_{ijk,l} - A_{ijl,k} = \tfrac{1}{2}\left(G_{ik}B_{jl} + G_{jk}B_{il} - G_{il}B_{jk} - G_{jl}B_{ik}\right), \tag{2.5.2}$$

$$\begin{aligned} R_{ijkl} =& \sum \left(A_{il}^{m}A_{mjk} - A_{ik}^{m}A_{mjl}\right) \\ &+ \tfrac{1}{2}\left(G_{ik}B_{jl} + G_{jl}B_{ik} - G_{il}B_{jk} - G_{jk}B_{il}\right), \end{aligned} \tag{2.5.3}$$

$$B_{ik,j} - B_{ij,k} = \sum \left(B_{jl}A_{ik}^{l} - B_{lk}A_{ij}^{l}\right). \tag{2.5.4}$$

For the proof we refer to [58], pp. 73-75.
While (2.5.3) is called an *integrability condition of Gauß type*, the other two systems are said to be of *Codazzi type*; this notion is analogous to the Euclidean theory.

Corollary. *By contraction the integrability conditions imply*

$$\sum A_{jk,l}^{l} = \tfrac{n}{2}\left(L_1 G_{jk} - B_{jk}\right), \tag{2.5.5}$$

$$R_{ik} = \sum A_{il}^{m}A_{mk}^{l} + \tfrac{n-2}{2}B_{ik} + \tfrac{n}{2}L_1 G_{ik}, \tag{2.5.6}$$

$$\sum B_{k,i}^{i} = nL_{1,k} + \sum B_{l}^{i}A_{ik}^{l}, \tag{2.5.7}$$

where L_1 is the affine mean curvature as before, and R_{ik} denote the local components of the Ricci tensor of (M, G).
From (2.5.6) we obtain, by another contraction, the so called

Equiaffine Theorema Egregium.

$$\kappa = J + L_1, \tag{2.5.8}$$

where

$$\kappa = \tfrac{1}{n(n-1)}\sum G^{ik}G^{jl}R_{ijkl}. \tag{2.5.9}$$

According to our notation in Riemannian geometry $R = n(n-1)\kappa$ is the scalar curvature and κ the normed scalar curvature of the metric G.

Corollary. *The form B can be expressed in terms of G, A and their derivatives:*

$$B_{jk} = (\kappa - J)\,G_{jk} - \tfrac{2}{n}\sum A_{jk,\,l}^{l}.$$

The proof follows from (2.5.5) and the Equiaffine Theorema Egregium; see [58], p.76.

2.6 Fundamental Theorem

As already stated the integrability conditions are necessary and sufficient for the integration of the structure equations of Gauß and Weingarten; one gets a local frame $\{e_1, ..., e_{n+1}\}$. Another integration gives the hypersurface.

Uniqueness Theorem. *Let x, $x^\sharp : M \to A^{n+1}$ be two non-degenerate hypersurfaces such that*

$$G = G^\sharp, \qquad A = A^\sharp.$$

Then x, $x^\sharp$ differ by a unimodular affine transformation; that means both hypersurfaces are equi-affinely equivalent.

Existence Theorem. *Let (M, G) be an n-dimensional semi-Riemannian manifold with metric G. Suppose that a symmetric cubic covariant tensor field*

$$A = \sum A_{ijk}\, \omega^i\, \omega^j \omega^k$$

is given on M. If G and A satisfy the apolarity condition and the integrability conditions then there exists a non-degenerate immersion $x : M \to A^{n+1}$ such that G and A are the Blaschke metric and the Fubini-Pick form for the immersion, respectively.

For a proof see [58], section 1.5.3.

Terminology. It is a consequence of the uniqueness Theorem that the pair (G, A) is a fundamental system of the hypersurface, that means one is able to determine all unimodular invariants of the hypersurface, and thus its geometry, from G and A. Because of the relations

$$\widetilde{\nabla} = \nabla + A, \qquad \nabla^* = \nabla - A$$

one can also consider the pairs $(G, \widetilde{\nabla})$ or (G, ∇^*) as fundamental systems.

Different versions of the Fundamental Theorem. There exist different versions of the Fundamental Theorem, namely to each fundamental system there is a modified version of the existence and uniqueness theorem. For proofs we refer to [88], chapter 4, and Theorem 3.5 in [86]. We will come back to the Fundamental Theorem in section 3.3.7 below.

2.7 Graph Immersions with Unimodular Normalization

Let $\Omega \subset \mathbb{R}^n$ be a domain and $x : M \to A^{n+1}$ be the graph of a strictly convex smooth function

$$x^{n+1} = f(x^1, \cdots, x^n), \qquad \text{where} \qquad (x^1, \cdots, x^n) \in \Omega \subset \mathbb{R}^n.$$

We choose the following unimodular affine frame field:

$$e_i = (0, ..., 0, 1, 0, ..., 0, \tfrac{\partial f}{\partial x^i}) \quad \text{for} \quad i = 1, ..., n \quad \text{and} \quad e_{n+1} = (0, \cdots, 0, 1).$$

Then the Blaschke metric is given by

$$G = \left[\det\left(\frac{\partial^2 f}{\partial x^j x^i}\right)\right]^{\frac{-1}{n+2}} \sum \frac{\partial^2 f}{\partial x^j x^i}\, dx^i\, dx^j,$$

and the affine conormal vector field U can be identified with

$$\left[\det\left(\frac{\partial^2 f}{\partial x^j x^i}\right)\right]^{\frac{-1}{n+2}} \left(-\frac{\partial f}{\partial x^1}, \ldots, -\frac{\partial f}{\partial x^n}, 1\right).$$

In the following we give some basic formulas with respect to the Blaschke metric; we will use them in later chapters.

The formula $\Delta U = -nL_1 U$ implies that $x(M)$ is a locally strongly convex hypersurface with constant affine mean curvature $L_1 =: L$ if and only if f satisfies the following PDE

$$\Delta\left\{\left[\det\left(\frac{\partial^2 f}{\partial x^i \partial x^j}\right)\right]^{\frac{-1}{n+2}}\right\} = -nL\left[\det\left(\frac{\partial^2 f}{\partial x^i \partial x^j}\right)\right]^{\frac{-1}{n+2}}, \tag{2.7.1}$$

where Δ denotes the Laplacian with respect to the Blaschke metric, which was defined in subsection 1.1.2. Recall the definition

$$\rho := \left[\det\left(\frac{\partial^2 f}{\partial x^i \partial x^j}\right)\right]^{\frac{-1}{n+2}}$$

from section 1.4. Then (2.7.1) gives

$$\Delta\rho = -nL\rho. \tag{2.7.2}$$

Note that, in terms of $x^1, ..., x^n$, we have $(\det(G_{kl}))^{\frac{1}{2}} = \frac{1}{\rho}$. By a direct calculation we get

$$\Delta = \sum G^{ij} \frac{\partial^2}{\partial x^i \partial x^j} - \frac{2}{\rho^2} \sum f^{ij} \frac{\partial \rho}{\partial x^j} \frac{\partial}{\partial x^i} + \frac{1}{\rho} \sum \frac{\partial f^{ij}}{\partial x^i} \frac{\partial}{\partial x^j}, \tag{2.7.3}$$

where (f^{ij}) denotes the inverse matrix of (f_{ij}) and $f_{ij} = \frac{\partial^2 f}{\partial x^i \partial x^j}$. Taking the differentiation of the equation $\sum f^{ik} f_{kj} = \delta^i_j$ one finds

$$\sum_{i,k} \frac{\partial f^{ik}}{\partial x^i} f_{kj} = -\sum_{i,k} f^{ik} \frac{\partial f_{kj}}{\partial x^i} = \frac{n+2}{\rho} \frac{\partial \rho}{\partial x^j}.$$

It follows that

$$\sum_i \frac{\partial f^{ik}}{\partial x^i} = \frac{n+2}{\rho} \sum_j f^{jk} \frac{\partial \rho}{\partial x^j}. \tag{2.7.4}$$

We insert (2.7.4) into (2.7.3) and obtain

$$\Delta = \frac{1}{\rho} \sum f^{ij} \frac{\partial^2}{\partial x^i \partial x^j} + \frac{n}{\rho^2} \sum f^{ij} \frac{\partial \rho}{\partial x^j} \frac{\partial}{\partial x^i}. \tag{2.7.5}$$

To find the affine normal Y and calculate the affine Weingarten tensor B, we let (see [27])

$$e_i^* = e_i, \quad 1 \le i \le n, \qquad e_{n+1}^* = e_{n+1} + \sum a_{n+1}^i e_i,$$

where e^*_{n+1} is in the affine normal direction. Since $\langle dU, e^*_{n+1}\rangle = 0$, the coefficients a^i_{n+1} are determined by

$$\sum a^j_{n+1} f_{ji} = \tfrac{\partial}{\partial x^i} \ln \rho.$$

It follows that

$$a^i_{n+1} = \sum f^{ji} \tfrac{\partial}{\partial x^j} \ln \rho,$$

and hence

$$e^*_{n+1} = e_{n+1} + \sum f^{ji} \tfrac{\partial}{\partial x^j} \ln \rho \cdot e_i.$$

Therefore

$$Y = H^{\frac{1}{n+2}} e^*_{n+1} = H^{\frac{1}{n+2}} \sum f^{ji} \tfrac{\partial}{\partial x^j} \ln \rho \cdot e_i + H^{\frac{1}{n+2}} e_{n+1},$$

where $H = \det\left(\frac{\partial^2 f}{\partial x^i \partial x^j}\right)$.

Let x denote the position vector of the hypersurface M. We have

$$dx = \sum w^\alpha e^*_\alpha,$$
$$de^*_\alpha = \sum w^\beta_\alpha e^*_\beta.$$

The w^α, w^β_α are the Maurer-Cartan forms of the unimodular affine group. We compute

$$\begin{aligned} w^i_{n+1} &= da^i_{n+1} - a^i_{n+1} d \ln \rho \\ &= \sum \left(\tfrac{\partial}{\partial x^j} \left(f^{ki} \tfrac{\partial}{\partial x^k} \ln \rho\right) - f^{ki} \tfrac{\partial}{\partial x^k} \ln \rho \cdot \tfrac{\partial}{\partial x^j} \ln \rho\right) w^j. \end{aligned}$$

Therefore the affine Weingarten tensor is

$$\begin{aligned} B_{ij} &= \sum \left(-\tfrac{\partial}{\partial x^i} \left(f^{lk} \tfrac{\partial}{\partial x^l} \ln \rho\right) + f^{lk} \tfrac{\partial}{\partial x^l} \ln \rho \tfrac{\partial}{\partial x^i} \ln \rho\right) f_{kj} \\ &= -\tfrac{1}{\rho} \tfrac{\partial^2 \rho}{\partial x^i \partial x^j} + \tfrac{2}{\rho^2} \tfrac{\partial \rho}{\partial x^i} \tfrac{\partial \rho}{\partial x^j} + \sum \tfrac{f^{kl}}{\rho} \tfrac{\partial \rho}{\partial x^l} \tfrac{\partial f_{ij}}{\partial x^k}. \end{aligned} \tag{2.7.6}$$

From section 1.4, recall the Legendre transformation relative to f, and denote again by Ω^* the Legendre transformation domain of f, i.e. $u : \Omega^* \to \mathbb{R}$ and

$$\Omega^* = \{(\xi_1(x), ..., \xi_n(x)) \mid x \in \Omega\}.$$

Considering a locally strongly convex graph, it is an advantage that we can express the basic formulas in terms of the x-coordinates as well in terms of the ξ-coordinates. In terms of the coordinates $(\xi_1, ..., \xi_n)$ and $u(\xi)$ the Blaschke metric is given by

$$G_{ij} = \rho \frac{\partial^2 u}{\partial \xi_i \partial \xi_j},$$

and $\left(\frac{\partial^2 u}{\partial \xi_i \partial \xi_j}\right)$ is the inverse matrix of $\left(\frac{\partial^2 f}{\partial x^i \partial x^j}\right)$. We have

$$\rho = \left[\det\left(\frac{\partial^2 u}{\partial \xi_i \partial \xi_j}\right)\right]^{\frac{1}{n+2}}, \tag{2.7.7}$$

$$\sqrt{\det(G_{kl})} = \rho^{n+1},$$

$$\Delta = \frac{1}{\sqrt{\det(G_{kl})}} \sum \frac{\partial}{\partial \xi_i}\left(G^{ij} \sqrt{\det(G_{kl})} \frac{\partial}{\partial \xi_j}\right)$$

(see [58], p.91). By a similar calculation as above we get

$$\Delta = \frac{1}{\rho} \sum u^{ij} \frac{\partial^2}{\partial \xi_i \partial \xi_j} - \frac{2}{\rho^2} \sum u^{ij} \frac{\partial \rho}{\partial \xi_j} \frac{\partial}{\partial \xi_i}. \tag{2.7.8}$$

2.8 Affine Spheres and Quadrics

As before we consider non-degenerate hypersurfaces with unimodular normalization.

2.8.1 *Affine hyperspheres*

For proofs and details we refer to section 2.1 in [58].

Definition. A non-degenerate hypersurface x in A^{n+1} is called an *affine hypersphere* if the affine normal line bundle has one of the following two properties:
(i) All affine normal lines meet at one point $c_0 \in A^{n+1}$; in this case x is called a *proper affine hypersphere* with center c_0.
(ii) All affine normal lines are parallel in A^{n+1}; in this case x is called an *improper affine hypersphere.*

Proposition. *Let M be a non-degenerate hypersurface in A^{n+1}.*
(a) *The following three properties (a.1)-(a.3) are equivalent:*
(a.1) *M is an affine hypersphere.*
(a.2) $B = L_1 \cdot G$.
(a.3) $B^\sharp = L_1 \cdot id$.
(b) *For an affine hypersphere we have $L_i = const$ for all $i = 1, ..., n$.*

Definition and Remark. Assume that x is locally strongly convex; that means that the Blaschke metric G is (positive) definite. In this case the affine Weingarten operator $B^\sharp$ has n real eigenvalues $\lambda_1, \lambda_2, \cdots, \lambda_n$, the *affine principal curvatures.* Then:

(i) The relation $B = L_1 \cdot G$ is equivalent to the equality of the affine principal curvatures:

$$\lambda_1 = \lambda_2 = \cdots = \lambda_n.$$

(ii) All affine principal curvatures are constant.

(iii) An affine hypersphere is called an *elliptic* affine hypersphere if $L_1 > 0$; it is called *hyperbolic* if $L_1 < 0$; it is called *parabolic* if $L_1 = 0$. Obviously the parabolic affine hyperspheres are exactly the improper affine hyperspheres.

(iv) For an elliptic affine hypersphere, the center is on the concave side of $x(M)$. For a hyperbolic affine hypersphere the center is on the convex side of $x(M)$. For a parabolic affine hypersphere we may consider the center to be at infinity.

(v) For a hypersurface in Euclidean space with Euclidean Weingarten operator S and mean curvature H the equation $S = H \cdot id$ implies $H = const$; if $H = 0$ we have a hyperplane, if $H \neq 0$ we have a sphere with curvature equal to $H > 0$.

In contrast to the Euclidean case the situation in the affine case is very complicated. There is an abundance of affine spheres, one knows many examples, but one is far from a classification. Only under strong additional assumptions there exist partial classifications. See e.g. the local classification of affine spheres with constant sectional curvature in [95], [96]; even under such strong additional conditions this classification is not yet finished.

Lemma. (i) *A non-degenerate hypersurface is an affine hypersphere if and only if the cubic form satisfies the covariant PDE*

$$A_{ijk,l} = A_{ijl,k}$$

with respect to the Levi-Civita connection.

(ii) *Both,* $\widetilde{\nabla} A$ *and* $\widetilde{\nabla}\widetilde{\nabla} A$*, the covariant derivatives in terms of the induced connection* $\widetilde{\nabla}$*, are totally symmetric if and only if* x *is a quadric (i.e.,* $A = 0$*) or* x *is an improper affine sphere (i.e.,* $B = 0$*).*

Proof. **(i)**: Apply (2.5.2). For **(ii)** see [11]. ∎

2.8.2 *Characterization of quadrics*

Theorem. (i) *Any hyperquadric is an affine hypersphere. The quadric has a center if* $L_1 \neq 0$.
(ii) *A non-degenerate hypersurface* x *is a quadric if and only if the cubic form* $A^\flat$ *vanishes identically on* M.

For a proof see [9] or section 1.4 in [58] (there we consider only locally strongly convex hypersurfaces). See also section 7 in [88] together with a clarifying Remark 2.2.(b) in [56]. In section 3.4 below we will generalize the foregoing Theorem.

Chapter 3

Local Relative Hypersurfaces

E. Müller was the first to extend the development of a unimodular hypersurface theory to a so called *relative hypersurface theory.* This concept is not only of interest from the geometric view point, but one can also apply it for the geometric solution of PDEs. Here we summarize the material necessary for our purposes. For details we refer to the two monographs [58] and [88], for a survey to [86]. For a unifying approach studying invariants that are independent of the choice of the normalization see [87].

In the following summary of the basic formulas we use the invariant and the local calculus; in this way we present the basic formulas from the affine hypersurface theories in three different terminologies, namely: in Chapter 2 Cartan's calculus together with a standard local calculus, in Chapter 3 the invariant calculus of Koszul and again a local description.

3.1 Hypersurfaces with Arbitrary Normalization

Recall section 2.1.1. In the following section A^{n+1} denotes a real affine space of dimension $n+1$. We identify geometric objects with respect to the general affine transformation group.

3.1.1 *Structure equations*

Normalizations. We consider a hypersurface as in (2.1.2). A *normalization* is a pair (U, z) where $U : M \to V^*$ is a conormal field as in section 2.3.3, and $z : M \to V$ is transversal to the hypersurface $x(M)$, both satisfying the relation $\langle U, z \rangle = 1$. A triple (x, U, z) is called a *normalized hypersurface.*

Structure equations. In analogy to Chapter 2 we can write down structure equations of Gauß and Weingarten type for a hypersurface x with normalization

(U, z):

$$\bar{\nabla}_v dx(w) = dx(\nabla_v w) + \mathfrak{h}(v, w)\, z,$$
$$dz(v) = dx(-Sv) + \theta(v)\, z.$$

We list the following elementary facts.

Properties of the coefficients. For a given triple (x, U, z) the coefficients in the structure equations have the following properties:

(i) ∇ is a torsion free connection on M; ∇ is called the *induced connection* of (x, U, z);
(ii) $\mathfrak{h}$ is a symmetric bilinear form over each tangent space;
(iii) S is a linear operator on each tangent space;
(iv) θ is a one-form;
(v) all coefficients are differentiable, they are invariant under the action of the general affine transformation group.

Lemma. *For a given hypersurface x and two different normalizations (U, z) and $(U^\sharp, z^\sharp)$, the induced bilinear forms $\mathfrak{h}$ and $\mathfrak{h}^\sharp$ in the structure equations satisfy $\mathfrak{h}^\sharp = q \cdot \mathfrak{h}$ for some non-zero factor $q \in C^\infty(M)$. As a consequence, the rank of $\mathfrak{h}$ does not depend on the choice of the transversal field z, it is a property of the hypersurface x itself.*

Non-degenerate hypersurfaces. x is called *non-degenerate* if, for an arbitrary normalization, $rank\, \mathfrak{h} = n$. If x is non-degenerate the class $\mathfrak{C} = \{\mathfrak{h}\}$ can be considered as a *conformal class of semi-Riemannian metrics*; in the definite case, by an appropriate orientation of the normalization, the class $\mathfrak{C}$ is positive definite and thus it is a class of Riemannian metrics.

Equivalence-Lemma. *For x the following properties are equivalent:*

(i) x is non-degenerate;
(ii) there exists a conormal field U such that, for an arbitrary frame $\{v_1, ..., v_n\}$:
$$rank\,(dU(v_1), ..., dU(v_n), U) = n + 1;$$
(iii) for any conormal field U and any frame the rank-condition in (ii) is satisfied.

As a consequence, for x non-degenerate, any conormal field defines an immersion

$$U : M \to V^*$$

with transversal vector field U. Thus one can write down structure equations of Gauß type for any conormal field U:

$$\bar{\nabla}_v dU(w) = dU(\nabla^*_v w) + \tfrac{1}{n-1} Ric^*(v, w)\, (-U).$$

$U : M \to V^*$ is called the *conormal indicatrix* of (x, U, z). One verifies:

Properties of the coefficients. For x non-degenerate with normalization (U, z) we have:

(i) The connection ∇^* is torsion free connection with symmetric Ricci tensor Ric^* on M; ∇^* is called *conormal connection*; it is well known that the Ricci-symmetry is equivalent to the fact that ∇^* admits a parallel volume form dV^*, i.e., $\nabla^* dV^* = 0$; the volume form is unique modulo a non-zero constant factor;
(ii) the induced invariants $\mathfrak{h}$ and ∇^* satisfy so called *Codazzi equations*, i.e., the covariant derivative $\nabla^* \mathfrak{h}$ is a totally symmetric cubic form;
(iii) all conormal connections ∇^* are projectively equivalent, i.e., they have the same unparametrized geodesics; the class $\mathfrak{P} = \{\nabla^*\}$ is projectively flat.

3.1.2 *Fundamental theorem for non-degenerate hypersurfaces*

Uniqueness Theorem. *Let (x, U, z) and $(x^\sharp, U^\sharp, z^\sharp)$ be non-degenerate hypersurfaces with the same parameter manifold: $x,\ x^\sharp : M \to A^{n+1}$. Assume that*

$$\mathfrak{h} = \mathfrak{h}^\sharp \quad and \quad \nabla^* = \nabla^{*\sharp}.$$

Then (x, U, z) and $(x^\sharp, U^\sharp, z^\sharp)$ are equivalent modulo a general affine transformation.

Existence Theorem. *On a connected, simply connected differentiable manifold M there are given:*

(i) a conformal class $\mathfrak{C} = \{\mathfrak{h}\}$ of semi-Riemannian metrics;
(ii) a projectively flat class $\mathfrak{P} = \{\nabla^*\}$ of torsion free, Ricci-symmetric connections;
(iii) there exists a pair $(\nabla^*, \mathfrak{h})$ such that they satisfy Codazzi equations.

Then there exists a non-degenerate hypersurface x such that $\mathfrak{C} = \{\mathfrak{h}\}$ is the class of induced bilinear forms, and $\mathfrak{P} = \{\nabla^*\}$ the induced class of conormal connections in the Gauß structure equations.

3.2 Hypersurfaces with Relative Normalization

From now on we consider non-degenerate hypersurfaces only. Following the geometric arguments in [87], one can restrict to the subclass of so called *relative normalizations*, namely: Two transversal fields $z,\ z^\sharp$ are called equivalent if they satisfy

$$\langle U, z \rangle = \langle U, z^\sharp \rangle$$

for one - and then for any - conormal field U. In each equivalence class there is exactly one representative Y of this class, satisfying the relations

$$\langle U, dY(v) \rangle = 0, \quad \langle dU(v), Y \rangle = 0, \quad \langle U, Y \rangle = 1.$$

We call Y a *relative normal*, the pair (U, Y) with $\langle U, Y \rangle = 1$ a *relative normalization*, and the triple (x, U, Y) with x non-degenerate a *relative hypersurface*. It is a consequence of the definition of a relative normalization that there is a bijective correspondence between conormals and relative normals. Moreover, for any two normalizations (U, z) and $(U, z^\sharp)$ of x in the same equivalence class, the symmetric bilinear forms coincide, $\mathfrak{h} = \mathfrak{h}^\sharp$, and also the conormals, $U = U^\sharp$, thus the conormal connections coincide: $\nabla^* = \nabla^{*\sharp}$. As the pair $(\nabla^*, \mathfrak{h})$ is a fundamental system for the triple (x, U, Y), it represents triples with equivalent transversal fields. The foregoing justifies our claim that one can restrict to the distinguished class of relative normalizations.

3.2.1 *Relative structure equations and basic invariants*

For a relative hypersurface (x, U, Y) the structure equations read:

Gauß equation for x $\qquad \bar{\nabla}_v dx(w) = dx(\nabla_v w) + \mathfrak{h}(v, w)\, Y,$

Weingarten equation $\qquad dY(v) = dx(-Sv),$

Gauß equation for U $\qquad \bar{\nabla}_v dU(w) = dU(\nabla^*_v w) + \frac{1}{n-1} Ric^*(v, w)\,(-U).$

For relative normalizations the geometric properties of most coefficients are better than in the case of arbitrary normalizations.

Properties of the coefficients. Let (x, U, Y) be a relative hypersurface. Then:

(i) The *induced connection* ∇ is torsion free and Ricci-symmetric.
(ii) The *relative shape operator* S is $\mathfrak{h}$-self-adjoint and satisfies

$$(n-1)S^\flat(v, w) := (n-1)\mathfrak{h}(Sv, w) = Ric^*(v, w).$$

Its trace gives the *relative mean curvature* $nL_1 := tr\, S$.
(iii) The triple $(\nabla, \mathfrak{h}, \nabla^*)$ is *conjugate*, that means it satisfies the following *generalization of the Ricci Lemma* in Riemannian geometry:

$$u\,\mathfrak{h}(v, w) = \mathfrak{h}(\nabla_u v, w) + \mathfrak{h}(v, \nabla^*_u w).$$

(iv) The Levi-Civita connection $\nabla(\mathfrak{h})$ of the non-degenerate *relative metric* $\mathfrak{h}$ satisfies

$$\nabla(\mathfrak{h}) = \tfrac{1}{2}(\nabla + \nabla^*).$$

It is a trivial consequence that any two of the three connections determine the third one.
(v) The covariant derivatives are totally symmetric (*Codazzi equations*) and satisfy

$$\nabla^* \mathfrak{h} = -\nabla \mathfrak{h}.$$

Cubic form and Tchebychev vector field. In analogy to the equiaffine theory we define:

$$A(v,w) := \nabla_v w \; - \; \nabla(\mathfrak{h})_v w \quad \text{and} \quad A^\flat(u,v,w) := \mathfrak{h}(u, A(v,w)).$$

$A^\flat$ is called the *relative cubic form.* In contrast to the unimodular theory this time the trace of A in general is non-zero, that is the apolarity condition is not valid. In fact: the apolarity condition characterizes the equiaffine normalization within the class of all relative normalizations [88].

We define the *relative Tchebychev form* $T^\flat$ as the trace of a linear mapping by

$$nT^\flat(v) := tr\{w \to A(w,v)\}$$

and the associated *relative Tchebychev vector field* T by

$$\mathfrak{h}(T,v) := T^\flat(v)$$

for all tangent fields v. One can easily show that the one-form $T^\flat$ is closed and thus T is the gradient of a potential function.

Relative structure equations in covariant local notation.
We rewrite the structure equations in terms of $\mathfrak{h}$-covariant differentiation and with affinely invariant coefficients as follows:

Gauß equation for x : $$x_{,ij} = \sum_k A_{ij}^k \, x_k + \mathfrak{h}_{ij} \, Y.$$

Weingarten equation : $$Y_j = - \sum_i S_j^i \, x_i.$$

Gauß equation for U : $$U_{,ij} = - \sum A_{ij}^k \, U_k - B_{ij} \, U,$$

where the relative Weingarten form satisfies $(n-1)B_{ij} = (n-1)\mathfrak{h}_{ik}S_j^k = R_{ij}^*$.

The relative support function. Let b be a fixed vector in V and U a relative conormal of x. The function $\Lambda : M \to \mathbb{R}$ defined by

$$\Lambda(p) := \langle U, b - x(p) \rangle, \quad p \in M,$$

is called the *relative support function* of (x, U, Y) with respect to the fixed point $b \in \mathbb{R}^{n+1}$.

In analogy to the Euclidean and the unimodular case the relative support function satisfies important PDEs; compare section 2.4.4:

$$\Lambda_{,ij} = - \sum A_{ij}^k \, \Lambda_k - \Lambda B_{ij} + \mathfrak{h}_{ij},$$

$$\Delta\Lambda + nT(\operatorname{grad}_{\mathfrak{h}} \Lambda) \; + \; nL_1\Lambda = n.$$

The relative Pick invariant. In analogy to the unimodular theory we define the *relative Pick invariant* by

$$J := \tfrac{1}{n(n-1)} \sum \mathfrak{h}^{il}\mathfrak{h}^{jm}\mathfrak{h}^{kr} A_{ijk} \, A_{lmr} = \tfrac{1}{n(n-1)} \, \|A\|^2,$$

where the tensor norm $\| \cdot \|$ is taken with respect to the relative metric $\mathfrak{h}$.

3.2.2 *Relative integrability conditions*

Like in the unimodular theory one derives the integrability conditions for relative hypersurfaces. They give information about relations between the invariants that appear in the relative structure equations. We are going to express these conditions in terms of the quadratic and cubic forms $\mathfrak{h}$, $S^\flat$, and $A^\flat$. Locally write $S_{ij} := S^\flat_{ij}$. We express the integrability conditions in terms of the metric $\mathfrak{h}$ and the cubic form A, in analogy to the classical approach in Blaschke's unimodular theory. We state:

3.2.3 *Classical version of the integrability conditions*

In covariant form the integrability conditions read:

$$A_{ijk,l} - A_{ijl,k} = \tfrac{1}{2}\left(\mathfrak{h}_{ik}S_{jl} + \mathfrak{h}_{jk}S_{il} - \mathfrak{h}_{il}S_{jk} - \mathfrak{h}_{jl}S_{ik}\right),$$

$$R_{ijkl} = \sum\left(A^m_{il}A_{mjk} - A^m_{ik}A_{mjl}\right) + \tfrac{1}{2}\left(\mathfrak{h}_{ik}S_{jl} + \mathfrak{h}_{jl}S_{ik} - \mathfrak{h}_{il}S_{jk} - \mathfrak{h}_{jk}S_{il}\right),$$

$$S_{ik,j} - S_{ij,k} = \sum\left(S_{jl}A^l_{ik} - S_{lk}A^l_{ij}\right).$$

By contraction, the integrability conditions imply:

(a) $\sum A^l_{jk,l} - nT_{,jk} = \frac{n}{2}\left(L_1\mathfrak{h}_{jk} - S_{jk}\right)$,

(b) $R(\mathfrak{h})_{ik} = \sum A^m_{il}A^l_{mk} - nT_lA^l_{ik} + \frac{1}{2}(n-2)S_{ik} + \frac{n}{2}L_1\,\mathfrak{h}_{ik}$,

(c) $\sum S^i_{k,i} = nL_{1,k} + \sum S^i_lA^l_{ik} - nS_{lk}T^l$,

where $R(\mathfrak{h})_{ik}$ denote the local components of the Ricci tensor $Ric(\mathfrak{h})$ on $(M,\mathfrak{h})$.

Relative Theorema Egregium.

$$\kappa(\mathfrak{h}) = J + L_1 - \tfrac{n}{n-1}\|T\|^2.$$

According to our notation in Riemannian geometry, $\kappa(\mathfrak{h})$ is the *normed relative scalar curvature* of the relative metric $\mathfrak{h}$.

3.2.4 *Classical version of the fundamental theorem*

Uniqueness Theorem. *Let (x,U,Y) and $(x^\sharp,U^\sharp,Y^\sharp)$ be non-degenerate hypersurfaces with the same parameter manifold: x, $x^\sharp : M \to A^{n+1}$. Assume that*

$$\mathfrak{h} = \mathfrak{h}^\sharp \quad \textit{and} \quad A = A^\sharp.$$

Then (x,U,Y) and $(x^\sharp,U^\sharp,Y^\sharp)$ are equivalent modulo a general affine transformation.

Existence Theorem. *On a connected, differentiable manifold M there are given:*

(i) a semi-Riemannian metric $\mathfrak{h}$;
(ii) a totally symmetric cubic form $A^\flat$

such that the integrability conditions in the classical version are satisfied. Then there exists a relative hypersurface (x, U, Y) such that $\mathfrak{h}$ is the relative metric and $A^\flat$ the relative cubic form.

3.3 Examples of Relative Geometries

There are several distinguished relative geometries that play an important role in affine hypersurface theory. In case we are going to use different normalizations of a centroaffine hypersurface x at the same time, we use marks to identify geometric objects of different relative geometries. For more details we refer to [88] and [8].

3.3.1 *The Euclidean normalization*

To identify Euclidean invariants of x we will use the mark *"E"*.
If V is equipped with a Euclidean inner product, we identify V and V^* according to the Theorem of Riesz. A hypersurface x is non-degenerate if and only if the Euclidean Weingarten operator $S(E)$ has maximal rank; this is equivalent to the fact that the Euclidean second fundamental form II has maximal rank. For a Euclidean normalization, according to the Gauß structure equations, $\mathfrak{h}(E) = II$ is the relative metric. Let μ denote the *Euclidean unit normal field* of the hypersurface x. At the same time, the Euclidean normal μ is the conormal field in this geometry, thus $(U(E), Y(E)) = (\mu, \mu)$ is a relative normalization. We denote by I, II, III, the three *Euclidean fundamental forms*, resp. The induced connection $\nabla(E)$ coincides with the Levi-Civita connection $\nabla(I)$ of the first fundamental form I, while $\nabla^*(E) = \nabla(III)$. We have the relations

$$2C^\flat(E) = \nabla^*(III)II = -\nabla(I)II$$

for covariant derivations of the second fundamental form, and the following expression for the Tchebychev form

$$T^\flat(E) = -\tfrac{1}{2n}\, d \ln |L_n(E)|;$$

here $L_n(E) = \det S(E)$ denotes the *Euclidean Gauß-Kronecker curvature.* The geometry induced from the Euclidean normalization is invariant under motions. The relative view point helps to unify methods of proof, in particular in extrinsic curvature theory; for more details see e.g. sections 6.1 and 6.4.2 in [88].

3.3.2 *The equiaffine (Blaschke) normalization*

To identify equiaffine invariants of x we will use the mark *"e"*.
As in Chapter 2, in the ambient space we fix a determinant form Det as volume form, the associated invariance group is the unimodular group. There is a (modulo orientation) unique normalization $(U(e), Y(e))$ within all relative normalizations,

characterized by the vanishing of its Tchebychev field: $T(e) = 0$ (*apolarity condition*). The transversal field $Y := Y(e)$ in this normalization historically is called the *affine normal field.* Nowadays the unimodular geometry is often called *Blaschke geometry*; this terminology should honour Blaschke's many contributions to this field (without ignoring important contributions by other authors). The geometry induced from the Blaschke normalization $(U(e), Y := Y(e))$ was sketched in Chapter 2, in particular we have $G(v, w) = \mathfrak{h}(e)(v, w)$. As stated before, this geometry is invariant under the unimodular transformation group (including parallel translations).

3.3.3 *The centroaffine normalization*

We will use the mark *"c"* to identify centroaffine invariants of x.
For a non-degenerate hypersurface it is well known that the set

$$\{p \in M \mid x(p) \text{ tangential}\}$$

is nowhere dense. Thus the position vector x is transversal almost everywhere; this property is independent of the choice of the origin. These facts and continuity arguments admit to restrict the investigations to the following situation:
One fixes the origin and this way identifies A^{n+1} with V, then we consider non-degenerate hypersurfaces with transversal position vector in V; as before we denote the position vector again by x. We call a hypersurface with always transversal position vector *centroaffine*; see pp. 15 and 37-39 in [73]. For such a hypersurface one can choose $Y(c) := \varepsilon x$ as relative normal where $\varepsilon = +1$ or $\varepsilon = -1$ is chosen appropriately (see below). The conormal $U(c)$ is oriented always such that

$$\langle U(c)\,,\, Y(c)\rangle \;=\; 1.$$

We recall the following definitions. A locally strongly convex, centroaffine hypersurface is called to be of

(i) *hyperbolic type*, if, for any point $x(p) \in V$, the origin $0 \in V$ and the hypersurface are on different sides of the affine tangent hyperplane $dx(T_pM)$; the *centroaffine normal vector field* then is given by $Y(c) \;:=\; +x$ (examples are hyperbolic affine hyperspheres in $\mathbb{R}^{n+1}$ centered at $0 \in \mathbb{R}^{n+1}$); according to the choice $Y(c) \;=\; x$ we modify the definition of the support functions and set $\Lambda := \langle U, x\rangle$;

(ii) *elliptic type*, if, for any point $x(p) \in V$, the origin $0 \in V$ and the hypersurface are on the same side of the affine tangent hyperplane $dx(T_pM)$; now the *centroaffine normal vector field* is given by $Y(c) \;:=\; -x$ (examples are elliptic affine hyperspheres in $\mathbb{R}^{n+1}$ centered at $0 \in \mathbb{R}^{n+1}$).

The different orientations of the centroaffine normalization on locally strongly convex hypersurfaces guarantee that the centroaffine metric, denoted by $\mathfrak{h}(c)$, is positive definite in both cases.

3.3.4 Graph immersions with Calabi metric

In section 2.7 we considered a graph immersion equipped with a Blaschke geometry. Calabi [19] considered a different normalization as follows (we use the mark *"ca"* to identify this geometry).
Let f be a strictly convex C^∞ function defined on a domain $\Omega \subset \mathbb{R}^n$, and consider the graph hypersurface

$$M := \{(x, f(x)) \mid x^{n+1} = f(x^1, ..., x^n), \quad (x^1, ..., x^n) \in \Omega\}.$$

For M we choose the canonical relative normalization given by
$Y := Y(ca) := (0, 0, ..., 1)$, then the conormal field $U := U(ca)$ is given by

$$U = (-f_1, ..., -f_n, 1);$$

here the reader should recall the notation for partial derivatives from section 1.1.1. We consider the Riemannian metric $\mathfrak{H} := \mathfrak{h}(ca)$ on M, defined by the Hessian of the graph function f:

$$\mathfrak{H} := \sum f_{ij} dx^i dx^j,$$

where as before $f_{ij} = \partial_i \partial_j f$. Then $\mathfrak{H}$ is the relative metric with respect to the relative normalization defined by $Y(ca)$. This metric is very natural for a convex graph; we call it the *Calabi metric*, in the literature one also finds the terminology *Hessian metric.* Using the conventions in a local notation, as before we denote the inverse matrix of the matrix (f_{ij}) by (f^{jk}), thus $f_{ij} \cdot f^{jk} = \delta^i_k$.
For some basic formulas see pp. 39-40 in [73], here we list some more.
Denote by $x = (x^1, ..., x^n, f(x^1, ..., x^n))$ the position vector of M. In covariant form, the *Gauß structure equation* reads

$$x_{,ij} = \sum A^k_{ij} x_k + f_{ij} Y. \tag{3.3.1}$$

One calculates the following relations (see e.g. [76]): The Levi-Civita connection with respect to the metric $\mathfrak{H}$ is determined by its Christofffel symbols

$$\Gamma^k_{ij} = \tfrac{1}{2} \sum f^{kl} f_{ijl},$$

and the Fubini-Pick tensor A_{ijk} and the Weingarten tensor satisfy

$$A_{ijk} = -\tfrac{1}{2} f_{ijk}, \quad B_{ij} = 0.$$

The relative Tchebychev vector field is given by

$$T := \tfrac{1}{n} \sum f^{ij} A^k_{ij} \partial_k.$$

Consequently, for the relative Pick invariant, we have:

$$J = \tfrac{1}{4n(n-1)} \sum f^{il} f^{jm} f^{kr} f_{ijk} f_{lmr}.$$

The *integrability conditions* and the Ricci tensor read

$$R_{ijkl} = \sum f^{mh} (A_{jkm} A_{hil} - A_{ikm} A_{hjl}), \tag{3.3.2}$$

$$A_{ijk,l} = A_{ijl,k}, \tag{3.3.3}$$

$$R_{ik} = \sum f^{mh} f^{lj} (A_{iml} A_{hjk} - A_{imk} A_{hlj}). \tag{3.3.4}$$

The scalar curvature satisfies

$$R = n(n-1)\kappa = \sum f^{ik} f^{mh} f^{lj} (A_{ihl}A_{mjk} - A_{hjl}A_{ikm}). \tag{3.3.5}$$

In terms of the Calabi metric $\mathfrak{H}$ of a graph hypersurface we calculate its Laplacian; for this we recall the auxiliary functions

$$\rho := [\det(f_{ij})]^{-\frac{1}{n+2}}, \qquad \Phi := \sum f^{ij} \tfrac{\partial \ln \rho}{\partial x^i} \tfrac{\partial \ln \rho}{\partial x^j}.$$

Moreover, we consider the Legendre transform function u of f (see section 1.4) and recall the involutionary character, using different coordinates x and ξ; in particular we have

$$\begin{aligned}\Delta &= \sum f^{ij} \tfrac{\partial^2}{\partial x^i \partial x^j} + \tfrac{n+2}{2\rho} \sum f^{ij} \tfrac{\partial \rho}{\partial x^j} \tfrac{\partial}{\partial x^i} \\ &= \sum u^{ij} \tfrac{\partial^2}{\partial \xi_i \partial \xi_j} - \tfrac{n+2}{2\rho} \sum u^{ij} \tfrac{\partial \rho}{\partial \xi_j} \tfrac{\partial}{\partial \xi_i}\,.\end{aligned}$$

Hence

$$\Delta\left(\sum (x^k)^2\right) = 2\sum f^{ii} + \tfrac{n+2}{2}\mathfrak{H}\left(\operatorname{grad}\ln\rho, \operatorname{grad}(\sum (x^k)^2)\right), \tag{3.3.6}$$

$$\Delta\left(\sum (\xi_k)^2\right) = 2\sum u^{ii} - \tfrac{n+2}{2}\mathfrak{H}\left(\operatorname{grad}\ln\rho, \operatorname{grad}(\sum (\xi_k)^2)\right), \tag{3.3.7}$$

and

$$\Delta \xi_k = -\tfrac{n+2}{2}\mathfrak{H}\left(\operatorname{grad}\ln\rho, \operatorname{grad}\xi_k\right). \tag{3.3.8}$$

3.3.5 *The family of conformal metrics $G^{(\alpha)}$*

The Calabi metric $\mathfrak{H}$ from section 3.3.4 generates a conformal class of metrics as follows:
For a fixed $\alpha \in \mathbb{R}$, set $G^{(\alpha)} := \rho^\alpha \mathfrak{H}$, here and later we call $G^{(\alpha)}$ an *α-metric*. Then, for any smooth function F, we have

$$\Delta F = \rho^\alpha \Delta^{(\alpha)} F - \tfrac{(n-2)\alpha}{2}\mathfrak{H}(\operatorname{grad}\ln\rho, \operatorname{grad}F), \tag{3.3.9}$$

where $\Delta^{(\alpha)}$ is the Laplacian with respect to the α-metric.

α-Ricci curvature. Denote by $R^{(\alpha)}_{ij}$ the Ricci curvature with respect to the α-metric, then (see [83])

$$\begin{aligned}R^{(\alpha)}_{ij} &= R_{ij} - \tfrac{n-2}{2}(\ln\rho^\alpha)_{,ij} + \tfrac{n-2}{4}(\ln\rho^\alpha)_{,i}(\ln\rho^\alpha)_{,j} \\ &\quad - \tfrac{1}{2}\left(\Delta(\ln\rho^\alpha) + \tfrac{n-2}{2} \parallel \operatorname{grad}\ln\rho^\alpha \parallel^2\right)\mathfrak{H}_{ij},\end{aligned} \tag{3.3.10}$$

here "," denotes the covariant derivation with respect to the Calabi metric $\mathfrak{H}$.

3.3.6 *Comparison of different relative geometries*

As already stated in section 3.1, all relative metrics define a conformal class $\mathfrak{C} = \{\mathfrak{h}\}$ and all conormal connections a projectively flat class $\mathfrak{P} = \{\nabla^*\}$ with torsion free, Ricci-symmetric connections. As the changes within a conformal and a projective class, resp., are well known, it is relatively easy to calculate the change from one relative geometry of x to another. This was done in chapters 5 and 6 of [88], for details we refer to this reference; see also [87]. Here we state only three relations of this type:

1. The two relative metrics $II := \mathfrak{h}(E)$ and $G := \mathfrak{h}(e)$ are related by

$$G = \mathfrak{h}(e) = |\det(S(E))|^{\frac{-1}{n+2}} \cdot \mathfrak{h}(E).$$

2. The two relative metrics $G := \mathfrak{h}(e)$ and $\mathfrak{h}(c)$ and the support function $\Lambda(e)$ are related by

$$\Lambda^{-1}(e) \cdot \mathfrak{h}(e) = \mathfrak{h}(c).$$

3. The centroaffine Tchebychev field $T(c)$ satisfies

$$T(c) = \tfrac{n+2}{2n} \operatorname{grad}_{\mathfrak{h}(c)} \ln \Lambda(e).$$

3.3.7 *Different versions of fundamental theorems*

In relative geometry one can state different versions of a Fundamental Theorem, using different fundamental systems $(\nabla^*, \mathfrak{h})$, or $(\nabla, \mathfrak{h})$, or $(A, \mathfrak{h})$, or even the conformal class $\mathfrak{C} = \{\mathfrak{h}\}$ together with the projectively flat class $\mathfrak{P} = \{\nabla^*\}$, [86]. Which version one will apply depends on the purpose.

The integrability conditions of the classical Blaschke version, based on the fundamental system $(A, \mathfrak{h})$, have a very complicated form; this is a disadvantage. But this version is useful for the application of subtle tools from Riemannian geometry, like maximum principles or the Laplacian Comparison Theorem.

The integrability conditions for the version in terms of $(\nabla^*, \mathfrak{h})$ are geometrically very transparent, depending on the fact that the connection ∇^* is projectively flat; the versions using the pairs $(\nabla, \mathfrak{h})$ or together the classes $\mathfrak{P}$ and $\mathfrak{C}$, resp., are modifications of the version using $(\nabla^*, \mathfrak{h})$. These versions lead to a much better understanding of the theory, based on the results in [28].

3.4 Gauge Invariance and Relative Geometry

To investigate the geometry of a given non-degenerate hypersurface, we have different possibilities for an appropriate choice of a normalization; even within the distinguished class of relative normalizations there are infinitely many possibilities. In general, the geometric invariants are different for different relative normalizations. Additionally, for most of the relative geometries the associated invariance

group is still unknown; see [8].
As a consequence, there was a systematic search for affine invariants that are independent of the relative normalization. This was done in [87]. For our purpose it is sufficient to state the following facts; for details see [87] and further references given there.

(i) The change from one relative normalization to another is equivalent to the *gauge transformations* of a related *Weyl geometry*; see [10]. For this reason, invariants that are independent of the relative normalization, are called *gauge invariants.*
(ii) Starting with a tentative relative normalization, one can construct the equiaffine normalization $(U(e), Y = Y(e))$ modulo a constant non-zero factor, and thus one can determine all equiaffine invariants modulo a constant non-zero factor: they only depend on the hypersurface itself (modulo a factor). This way one proves:
All equiaffine invariants are gauge invariants (some modulo a factor).
This implies that properties of important classes of hypersurfaces in the unimodular theory like affine spheres, extremal hypersurfaces, etc., do not depend on the special choice of the normalization, the definitions of the classes are gauge invariant.
(iii) For any two relative normalizations (U, Y) and $(U^{\sharp}, Y^{\sharp})$ of x one has
$$\Lambda^{-1} \cdot \mathfrak{h} = \Lambda^{\sharp\,-1} \cdot \mathfrak{h}^{\sharp}.$$
As the centroaffine support function satisfies $\Lambda(c) = 1$, we see that
$$\mathfrak{h}(c) = \Lambda^{-1} \cdot \mathfrak{h}$$
for any relative normalization. From this *the centroaffine metric and its intrinsic geometry are gauge invariant*; one can finally prove that *this is true for all centroaffine invariants.*
(iv) The *Calabi geometry* from section 3.3.4 is a gauge invariant geometry. Consider an arbitrary relative normalization (U, Y) and the Calabi normalization
$$(U(ca), Y(ca)) = ((-\partial_1 f, ..., -\partial_n f, 1), (0, ..., 0, 1)).$$
Then
$$U(ca) = qU$$
where $0 < q \in C^{\infty}(M)$ and
$$\langle(-\partial_1 f, ..., -\partial_n f, 1), Y\rangle = q.$$
Therefore one can easily construct $(U(ca), Y(ca))$ from an arbitrary normalization (U, Y). Thus the *Calabi geometry* is gauge invariant.

We give two examples of gauge invariants, i.e., we express them in terms of an arbitrary relative normalization.

The traceless tensor field $\widetilde{A}$ and the characterization of quadrics. On a non-degenerate hypersurface, let U be an arbitrary conormal field; from U one can define the corresponding metric $\mathfrak{h}$ and the projectively flat connection ∇^*, and from this A and T. Define the symmetric (1,2) tensor field $\widetilde{A}$ as traceless part of A:

$$\widetilde{A}(v,w) := A(v,w) - \tfrac{n}{n+2}(T^\flat(v)w + T^\flat(w)v + \mathfrak{h}(v,w)T);$$

then

(a) $\widetilde{A}$ is a gauge invariant;
(b) $\widetilde{A} = A(e)$, thus $\widetilde{A} = 0$ if and only if the hypersurface is a hyperquadric;
(c) we calculate

$$\|\widetilde{A}\|^2 = \|A\|^2 - \tfrac{3n^2}{n+2}\|T\|^2.$$

Here the norms are defined via the relative metric used. In case that x is locally strongly convex and the orientation of the normalization is appropriate, any relative metric is positive definite. Then the foregoing identity allows to estimate $\|A\|^2$ in terms of $\|T\|^2$.

Affine spheres. Consider a non-degenerate, centroaffine hypersurface. We define

$$\widetilde{T}^\flat := T^\flat + \tfrac{n+2}{2n}\, d\ln\Lambda$$

and $\widetilde{T}$ implicitly by $\mathfrak{h}(c)(\widetilde{T},v) := \widetilde{T}^\flat(v)$. We state:

(i) $\widetilde{T} = 0$ if and only if x is a proper affine sphere.
(ii) $\widetilde{T} = T(c) = \frac{n+2}{2n}\,\mathrm{grad}_{\mathfrak{h}(c)} \ln \Lambda(e)$; compare section 3.3.6 above.

Completeness conditions. Below we are going to consider different types of completeness conditions. It is a consequence of the foregoing facts that all such completeness conditions are gauge invariant conditions; see [87].

Euler-Lagrange equations. In the following we will investigate classes of hypersurfaces that satisfy Euler-Lagrange equations of certain variational problems. One verifies that such Euler-Lagrange equations are again gauge invariant relations; see [87].

Chapter 4

The Theorem of Jörgens-Calabi-Pogorelov

In this chapter we are going to use geometric tools for the solution of certain types of Monge-Ampère equations. For this interplay of global affine differential geometry and PDEs we use the terminology *geometric modelling technique.* E. Calabi, A.V. Pogorelov, S.Y. Cheng, S.T. Yau, N.S. Trudinger, X.J. Wang, A.-M. Li and other authors (see e.g. [19], [20], [25], [54], [55], [60], [61], [62], [76], [91], [92]) developed the following method of geometric modelling:
One interprets the unknown convex function in the PDE as locally strongly convex global graph and chooses an appropriate relative normalization. The aim is to use special induced geometric structures of relative hypersurface theory to express the given PDE in terms of geometric invariants, while global assumptions for the PDE are interpreted in terms of appropriate geometric completeness conditions. For the solution of the PDE considered, it is crucial to estimate appropriate geometric invariants that are related to the problem.
In this chapter we will give typical examples of this geometric modelling, studying PDEs that are related to affine spheres and some generalizations of such PDEs.
For the convenience of the reader we recall the notation for the unimodular theory from Chapter 2, and Calabi's relative normalization for a graph from section 3.3.4. In sections 4.1-4.3 we summarize tools for the proofs of our results in sections 4.4-4.6.

4.1 Affine Hyperspheres and their PDEs

We consider affine hyperspheres and derive their PDEs. We will treat the two types separately: improper and proper affine hyperspheres. We characterize affine hyperspheres in terms of their PDEs. As the global results in this monograph only concern locally strongly convex hypersurfaces, in proofs we will restrict to this case.

4.1.1 *Improper affine hyperspheres*

Let x be a locally strongly convex parabolic affine hypersphere in A^{n+1}. By a unimodular affine transformation we can assume that the affine normal Y is given by

$$Y = (0, \cdots, 0, 1),$$

and that $x(M)$ is locally described in terms of a strictly convex function f, defined on a domain $\Omega \subset \mathbb{R}^n$:

$$x^{n+1} = f(x^1, \cdots, x^n).$$

We choose a local unimodular affine frame field for x as follows:

$$e_i = (0, ..., 0, 1, 0, ..., 0, \partial_i f), \quad i = 1, ..., n, \quad e_{n+1} = (0, ..., 0, 1). \tag{4.1.1}$$

The Gauß structure equations for x from section 2.4.3, and the relations

$$\langle U, e_{n+1} \rangle = 1, \quad \langle U, e_i \rangle = 0, \quad \text{for} \quad i = 1, ..., n,$$

give (see section 3.3.4)

$$h_{ij} = \partial_j \partial_i f.$$

Equation (2.3.2) and the apolarity condition imply $Y = |H|^{\frac{1}{n+2}} e_{n+1} = e_{n+1}$ and

$$H = \det(\partial_j \partial_i f) = 1. \tag{4.1.2}$$

Conversely, suppose that M is locally given by the strictly convex graph

$$x^{n+1} = f(x^1, \cdots, x^n),$$

and that f satisfies the PDE (4.1.2). Considering the frame field (4.1.1), we have $d \ln H = 0$ and $de_{n+1} = 0$. Hence

$$\omega_{n+1}^{n+1} = 0.$$

It follows that $e_{n+1} = (0, \cdots, 0, 1)$ is the affine normal vector Y at each point of $x(M)$. This shows that x is a parabolic affine hypersphere.

Theorem. *x is a parabolic affine hypersphere with constant affine normal vector $(0, \cdots, 0, 1)$ if and only if f satisfies the PDE (4.1.2) of Monge-Ampère type.*

4.1.2 *Proper affine hyperspheres*

Let x be an elliptic or hyperbolic affine hypersphere and assume that x locally is given as a graph of a strictly convex C^∞-function on a domain $\Omega \subset \mathbb{R}^n$:

$$x^{n+1} = f(x^1, \cdots, x^n), \qquad (x^1, \cdots, x^n) \in \Omega.$$

Consider the Legendre transformation from section 1.4:

$$F : \Omega \to \mathbb{R}^n \quad \text{where} \quad (x^1, \cdots, x^n) \mapsto (\xi_1, \cdots, \xi_n) \tag{4.1.3}$$

and

$$\xi_i := \partial_i f = \frac{\partial f}{\partial x^i}, \qquad i = 1, 2, \cdots, n.$$

When Ω is convex, $F : \Omega \to F(\Omega)$ is a diffeomorphism. Then the hypersurface can be represented in terms of $(\xi_1, \xi_2, \cdots, \xi_n)$ as follows:

$$x = \left(x^1, \cdots, x^n, f\left(x^1, \cdots, x^n\right)\right) = \left(\frac{\partial u}{\partial \xi_1}, \cdots, \frac{\partial u}{\partial \xi_n}, -u + \sum \xi_i \frac{\partial u}{\partial \xi_i}\right). \tag{4.1.4}$$

The affine normal satisfies (2.3.1) and (see p. 91 in [58])

$$Y = \frac{1}{n}\Delta x = \left(\rho^{-2}\frac{\partial \rho}{\partial \xi_1}, \cdots, \rho^{-2}\frac{\partial \rho}{\partial \xi_n}, \rho^{-1} + \rho^{-2}\sum \xi_i \frac{\partial \rho}{\partial \xi_i}\right), \tag{4.1.5}$$

where ρ is defined in section 1.4.

The relations (4.1.4) and (4.1.5) imply that the necessary and sufficient condition for the equality $Y = -L_1 x$ is given by the relation

$$\rho = \frac{1}{L_1 u}, \quad L_1 = const \neq 0, \tag{4.1.6}$$

i.e.,

$$\det\left(\frac{\partial^2 u}{\partial \xi_i \partial \xi_j}\right) = (L_1 u)^{-n-2} \tag{4.1.6.a}$$

where L_1 is a constant. We summarize the foregoing results:

Theorem. *Let x be an immersed hypersurface in A^{n+1} which locally is given as graph of a strictly convex C^∞-function $x^{n+1} = f\left(x^1, \cdots, x^n\right)$ over a convex domain. Then x is an elliptic or hyperbolic affine hypersphere with center at the origin if and only if the Legendre transform function u of f satisfies the PDE of Monge-Ampère type (4.1.6.a).*

As a consequence of the two foregoing theorems we can state the following:
Any solution of the Monge-Ampère equation (4.1.2) locally defines an improper affine sphere given as the graph of this solution. Any solution of the Monge-Ampère equation (4.1.6.a) similarly locally defines a proper affine sphere.

The characterization of affine spheres in terms of their PDEs explains our statement at the end of section 2.8, namely that both classes of affine spheres are very large.

4.1.3 *The Pick invariant on affine hyperspheres*

We recall a well known inequality for the Laplacian of the Pick invariant on affine hyperspheres. For $n = 2$ it first was obtained by W. Blaschke [9]. For higher dimensional affine spheres it was obtained by E. Calabi [19] in the case of parabolic affine hyperspheres, and for arbitrary affine hyperspheres by R. Schneider [82] (with a minor misprint of a constant) and also by Cheng and Yau [25]. U. Simon calculated ΔJ for arbitrary non-degenerate hypersurfaces and applied his formula to get some new characterization of ellipsoids (cf. [84], [85]). There are extension to the cubic form in relative geometry [56] and for Codazzi tensors of arbitrary order in Riemannian geometry [67], [68].

Lemma. *On a locally strongly convex affine hypersphere we have:*

$$\frac{n(n-1)}{2}\Delta J \geq \|\nabla A\|^2 + n(n-1)(n+1)J(J + L_1), \tag{4.1.7}$$

here ∇A denotes the covariant derivative of the Fubini-Pick form A.

4.2 Completeness in Affine Geometry

In affine differential geometry there are different notions of *completeness* for a locally strongly convex hypersurface x. Principally, one can consider the *completeness of any relative metric.* In subsection 4.2.1 we list the completeness notions that are of importance for our investigations. In later sections we will study relations between different notions of completeness.

4.2.1 *Affine completeness and Euclidean completeness*

Definition. (1) *Affine completeness* of M, that is the completeness of the Blaschke metric G (sections 2.2.1 and 2.7) on M;
(2) *Calabi completeness* of M, that is the completeness of the Calabi metric $\mathfrak{H}$ (section 3.3.4) on M;
(3) *Euclidean completeness* of M is the completeness of the Riemannian metric induced from an arbitrary Euclidean metric on the affine space A^{n+1}.

Lemma. [81]. *The notion of Euclidean completeness on M is independent of the choice of a Euclidean metric on A^{n+1}.*

Proof. Consider two inner products on V, denoted by $\langle\ ,\ \rangle$ and $\langle\langle\ ,\ \rangle\rangle$; they define two Euclidean metrics on A^{n+1}. Let $\eta_1, \eta_2, \cdots, \eta_{n+1}$ and $\bar{\eta}_1, \bar{\eta}_2, \cdots, \bar{\eta}_{n+1}$ be orthonormal bases in V relative to $\langle\ ,\ \rangle$ and $\langle\langle\ ,\ \rangle\rangle$, respectively, related by

$$\eta_\alpha = \sum C_\alpha^\beta \, \bar{\eta}_\beta$$

where $C = (C_\alpha^\beta) \in GL(n+1, \mathbb{R})$. The Euclidean structures of V induce Euclidean metrics on M; we can write them in the form

$$ds^2 = \left(dx^1, dx^2, \cdots, dx^{n+1}\right) \begin{pmatrix} dx^1 \\ dx^2 \\ \cdot \\ \cdot \\ \cdot \\ dx^{n+1} \end{pmatrix} := (dx)^\tau \cdot (dx);$$

$$d\bar{s}^2 = (C \cdot dx)^\tau \cdot C \cdot dx = dx^\tau \cdot C^\tau C \cdot dx;$$

here we use an obvious matrix notation, and C^τ denotes the transposed matrix of C. Let μ and λ denote the largest and smallest eigenvalues of the product matrix $C^\tau C$, respectively; they are positive. Then

$$\lambda ds^2 \leq d\bar{s}^2 \leq \mu ds^2.$$

This means that a curve in M has infinite length in one metric if and only if its length is infinite in the other metric. Thus the two notions of Euclidean completeness, induced on M from different Euclidean structures on A^{n+1}, are equivalent. This justifies to use the notion of *Euclidean completeness* in affine hypersurface theory. ∎

Remarks and Example.
(i) A global graph in $\mathbb{R}^{n+1}$ over $\mathbb{R}^n$ is Euclidean complete.

(ii) Generally, the notions of *affine completeness* and *Euclidean completeness* on M are *not* equivalent. R. Schneider gave the following example [81]:
Consider the graph given by

$$x^3 = f\left(x^1, x^2\right) = \tfrac{1}{2}\left(\tfrac{1}{x^1} + (x^2)^2\right),$$
$$x(M) = \left\{\left(x^1, x^2, f(x^1, x^2)\right) \mid 0 < x^1 < \infty, \quad -\infty < x^2 < \infty\right\}.$$

It is not difficult to check that x is locally strongly convex and Euclidean complete. On the other hand, the Blaschke metric G of x is given by

$$G_{11} = (x^1)^{-\frac{9}{4}}, \quad G_{12} = G_{21} = 0, \quad G_{22} = (x^1)^{\frac{3}{4}}.$$

On M, consider the curve

$$x^1(t) = t, \quad x^2(t) = 0, \quad 1 \leq t < \infty.$$

Its affine arc length is

$$l = \int_1^\infty \sqrt{G_{11}(t)}\, dt = \int_1^\infty t^{-\frac{9}{8}}\, dt < \infty.$$

This shows that M is not affine complete.

(iii) In [92] Trudinger and Wang proved the following

Theorem. *Let $n \geq 2$. If M is an affine complete, locally uniformly convex hypersurface in $\mathbb{R}^{n+1}$, then M is Euclidean complete.*

(iv) In sections 5.9 and 6.1 below, we will study relations between the different notions of completeness under additional assumptions.

4.2.2 *The Cheng-Yau criterion for affine completeness*

It follows from Schneider's example that one needs additional assumptions to prove that Euclidean completeness implies affine completeness. The first result of this type is due to Cheng and Yau [25]; they proved that for an affine hypersphere the Euclidean completeness implies the affine completeness. In section 6.1 we will extend this result for surfaces with constant affine mean curvature. Another related result was proved in [98].

Theorem. *Every Euclidean complete affine hypersphere is affine complete.*

We state a generalization of the result of Cheng and Yau; it is obvious that their Theorem is a corollary of the following criterion.

Completeness Criterion. *Let M be a locally strongly convex, Euclidean complete hypersurface in A^{n+1}. If there is a constant $N > 0$ such that the G-norm of the Weingarten form B is bounded from above:*

$$\| B \|_G \leq N, \tag{4.2.1}$$

then M is affine complete.

To prove the Completeness Criterion we will apply the Estimate Lemma below. The proof of the Lemma follows in the next subsection.

Special choice of the coordinate system. We consider a non-compact, Euclidean complete, locally strongly convex hypersurface $x : M \rightarrow A^{n+1}$. From Hadamard's Theorem in section 1.2 the hypersurface x is the graph of a strictly convex function f:

$$x^{n+1} = f\left(x^1, \cdots, x^n\right),$$

defined on a convex domain $\Omega \subset \mathbb{R}^n$. Hence x is globally strongly convex.

Claim: We may assume that the hyperplane $x^{n+1} = 0$ is the tangent hyperplane of x at some point $x_0 = (\dot{x}^1, \dot{x}^2, \cdots, \dot{x}^n) \in M$, and that x_0 has the coordinates $(0, \cdots, 0)$. This can easily be seen from the following:
For f given as above, define

$$\widetilde{x}^i = x^i - \dot{x}^i, \;\; 1 \leq i \leq n,$$

$$\widetilde{f}(\widetilde{x}) = f\left(x^1, \cdots, x^n\right) - \sum \tfrac{\partial f}{\partial x^i}\left(\dot{x}^1, \cdots, \dot{x}^n\right)\left(x^i - \dot{x}^i\right) - f\left(\dot{x}^1, \cdots, \dot{x}^n\right),$$

for any $\left(x^1, \cdots, x^n\right) \in \Omega$. Then the graph of $\widetilde{f}(\widetilde{x})$ has the required properties. Since the above transformation is affine, our claim is proved.

Remark. With respect to this special choice of the coordinate system we have $f \geq 0$; for any number $C > 0$, denote the *section* (for the definition see section 1.1.4 above or section 3.1 in [37])

$$S_f(0, C) := \left\{p \in \Omega \mid x^{n+1} = f\left(x^1, \cdots, x^n\right) < C\right\}.$$

Estimate Lemma. *Consider a non-compact, Euclidean complete, locally strongly convex hypersurface x with graph function f and with the special choice of the coordinates just described. Assume that there exists a real positive N such that the norm of the Weingarten form is bounded above as in (4.2.1). Then:*

(i) *There exists $N^\sharp \geq N$ such that the Laplacian satisfies the following estimate:*

$$\tfrac{|\Delta f|}{1+f} \leq N^\sharp. \tag{4.2.2}$$

(ii) *There exists a positive real Q such that f satisfies the following gradient estimate:*

$$\tfrac{\|\text{grad } f\|}{1+f} \leq Q. \tag{4.2.3}$$

Proof of the Completeness Criterion. We apply the Estimate Lemma. Let $p_0 \in M$. For any unit speed geodesic σ starting at p_0

$$\sigma : [0, S] \rightarrow M$$

we have

$$\tfrac{df}{ds} \le \|\operatorname{grad} f\| \le Q(1+f).$$

It follows that

$$S \ge \tfrac{1}{Q} \int_0^{x^{n+1}(\sigma(S))} \tfrac{df}{1+f} \,. \tag{4.2.4}$$

Since

$$\int_0^{\infty} \tfrac{df}{1+f} = \infty$$

and $f : \Omega \to \mathbb{R}$ is proper in the topological sense (i.e., the inverse image of any compact set is compact), (4.2.4) implies the affine completeness of (M, G). ∎

4.2.3 *Proof of the Estimate Lemma*

(i) Suppose that there is constant $N > 0$ such that (4.2.1) is satisfied. Consider the function

$$\varphi := (C - f)\, \tfrac{|\Delta f|}{1+f} \tag{4.2.5}$$

defined on $S_f(0, C)$. Obviously φ attains its supremum at some interior point x^* of $S_f(0, C)$. Without loss of generality we may assume that $|\Delta f| \neq 0$ at x^*; then $\operatorname{grad} \varphi = 0$ at x^*. Choose a local orthonormal frame field $\{e_1, \cdots, e_n\}$ of the Blaschke metric on M such that, at x^*:

$$f_{,1} = \|\operatorname{grad} f\| \quad \text{and} \quad f_{,i} = 0 \quad \text{for} \quad 2 \le i \le n,$$

where $f_{,i}$ satisfies $df = \sum f_{,i}\, \omega^i$. Then, at x^*,

$$\tfrac{-f_{,1} |\Delta f|}{1+f} - (C - f)\, \tfrac{f_{,1} |\Delta f|}{(1+f)^2} + (C - f)\, \tfrac{|\Delta f|_{,1}}{1+f} = 0.$$

It follows that

$$\tfrac{f_{,1} |\Delta f|}{1+f} \le |\Delta f|_{,1}.$$

Taking the $(n+1)$-st component of the identity

$$(\Delta x)_{,i} = nY_{,i} = -n \sum_j B_{ij}\, x_{,j},$$

we get

$$(\Delta f)_{,i} = -n \sum_j B_{ij} f_{,j} \quad \text{and} \quad |\Delta f|_{,1} \le nN f_{,1}.$$

Hence, from the assumption, at x^*,

$$\tfrac{f_{,1} |\Delta f|}{1+f} \le nN f_{,1}.$$

Note that $f(x_0) = \inf_{S_f(0,C)} f(x)$.
In the case $x^* \neq x_0$ we have $f_{,1} = \|\operatorname{grad} f\| > 0$. It follows that

$$\tfrac{|\Delta f|}{1+f} \le nN \quad \text{and} \quad \varphi \le (C - f)\, nN \le CnN. \tag{4.2.6}$$

(4.2.6) holds at x^* where φ attains its supremum.
In the case $x^* = x_0$ we have

$$\varphi \leq C \tfrac{|\Delta f|}{1+f} \,|_{x_0} = C\,|\Delta f(x_0)|.$$

Let $N^\sharp := \max\{nN,\ |\Delta f(x_0)|\}$. Then, at any point of $S_f(0,C)$:

$$\tfrac{|\Delta f|}{1+f} \leq \tfrac{CN^\sharp}{C-f}.$$

For $C \to \infty$ we arrive at the asserted estimate in **(i)**.

(ii) Now we are going to prove the gradient estimate for f. Consider the function

$$\psi := \exp\left\{\tfrac{-m}{C-f}\right\} \tfrac{\|\mathrm{grad}\, f\|^2}{(1+f)^2}$$

defined on $S_f(0,C)$, where m is a positive constant to be determined later. Clearly, ψ attains its supremum at some interior point x^* of $S_f(0,C)$. We can assume that $\|\mathrm{grad}\, f\| > 0$ at x^*. Choose a local orthonormal frame field $\{e_1, \cdots, e_n\}$ on M such that, at x^*,

$$f_{,1} = \|\mathrm{grad}\, f\|, \qquad f_{,i} = 0 \quad (2 \leq i \leq n).$$

Then, at x^*,

$$\psi_{,i} = 0 \quad \text{and} \quad \sum_i \psi_{,ii} \leq 0.$$

We calculate both expressions explicitly. At x^*, we finally get:

$$\tfrac{-m\, f_{,i}}{(C-f)^2} \sum_j (f_{,j})^2 - \tfrac{2f_{,i}}{1+f} \sum_j (f_{,j})^2 + 2\sum_j f_{,j}\, f_{,ij} = 0, \tag{4.2.7}$$

$$\begin{aligned} &-2\left[\tfrac{m}{(C-f)^2} + \tfrac{2}{1+f}\right](f_{,1})^2 f_{,11} + \left[\tfrac{-2m}{(C-f)^3} + \tfrac{2}{(1+f)^2}\right](f_{,1})^4 \\ &-\left[\tfrac{m}{(C-f)^2} + \tfrac{2}{1+f}\right](\Delta f)(f_{,1})^2 + 2\sum (f_{,ij})^2 + 2\sum f_{,j}\, f_{,jii} \leq 0. \end{aligned} \tag{4.2.8}$$

We insert (4.2.7) into (4.2.8) and obey the inequality $\frac{2}{(1+f)^2} > 0$; we get

$$\begin{aligned} &-\left[\tfrac{m}{(C-f)^2} + \tfrac{2}{1+f}\right]^2 (f_{,1})^4 - \tfrac{2m}{(C-f)^3}(f_{,1})^4 + 2\sum (f_{,ij})^2 \\ &-\left[\tfrac{m}{(C-f)^2} + \tfrac{2}{1+f}\right](\Delta f)(f_{,1})^2 + 2\sum f_{,j}\, f_{,jii} \leq 0. \end{aligned} \tag{4.2.9}$$

Let us now compute the terms $f_{,ij}$ and $f_{,jii}$. An application of the Ricci identities shows:

$$\sum f_{,j} f_{,jii} = \sum f_{,j}\, (\Delta f)_{,j} + \sum R_{ij}\, f_{,i}\, f_{,j}\,.$$

We apply the integrability conditions and insert the Ricci tensor into the foregoing expression

$$R_{ij} = \sum A_{mli} A_{mlj} + \tfrac{n-2}{2} B_{ij} + \tfrac{n}{2} L_1 \delta_{ij},$$

moreover we use the Weingarten structure equation

$$(\Delta f)_{,j} = -n \sum B_{ij}\, f_{,i}$$

to obtain

$$\sum f_{,j}\, f_{,jii} = \sum (A_{m\,l1})^2\, (f_{,1})^2 - \tfrac{n+2}{2}\, B_{11}\, (f_{,1})^2 + \tfrac{n}{2} L_1 (f_{,1})^2. \tag{4.2.10}$$

We take the $(n+1)$-st component of $x_{,ij} = \sum A_{ijk}\, x_{,k} + \frac{\Delta x}{n}\, \delta_{ij}$ and get

$$f_{,ij} = A_{ij1}\, f_{,1} + \tfrac{\Delta f}{n}\, \delta_{ij},$$

$$\sum (f_{,ij})^2 = \sum \left(A_{ij1}\, f_{,1} + \tfrac{\Delta f}{n}\, \delta_{ij}\right)^2 = \sum (A_{ij1} f_{,1})^2 + \tfrac{1}{n} (\Delta f)^2. \tag{4.2.11}$$

Combination of (4.2.10) and (4.2.11) gives

$$\sum f_{,j}\, f_{,jii} = \sum (f_{,ij})^2 + \tfrac{n}{2} L_1\, (f_{,1})^2 - \tfrac{n+2}{2}\, B_{11}\, (f_{,1})^2 - \tfrac{1}{n} (\Delta f)^2. \tag{4.2.12}$$

We apply the inequality of Schwarz and obtain:

$$\begin{aligned} \sum (f_{,ij})^2 &\geq (f_{,11})^2 + \sum_{i>1} (f_{,ii})^2 \geq (f_{,11})^2 + \tfrac{1}{n-1} \left(\sum_{i>1} f_{,ii} \right)^2 \\ &= \tfrac{n}{n-1} (f_{,11})^2 + \tfrac{(\Delta f)^2}{n-1} - \tfrac{2 f_{,11} \cdot \Delta f}{n-1} \\ &\geq \left(\tfrac{n}{n-1} - \delta \right) (f_{,11})^2 - \tfrac{1-\delta(n-1)}{\delta\, (n-1)^2} (\Delta f)^2 \end{aligned} \tag{4.2.13}$$

for any $\delta > 0$. Next we insert (4.2.12) and (4.2.13) into (4.2.9), together with (4.2.7) this implies the following inequality:

$$\left(\left(\tfrac{1}{n-1} - \delta \right) \left[\tfrac{m}{(C-f)^2} + \tfrac{2}{1+f} \right]^2 - \tfrac{2m}{(C-f)^3} \right) (f_{,1})^4 - \left[\tfrac{m}{(C-f)^2} + \tfrac{2}{1+f} \right] (\Delta f) \cdot (f_{,1})^2$$

$$+ (nL_1 - (n+2) B_{11}) (f_{,1})^2 - \left[\tfrac{4-4(n-1)\delta}{\delta (n-1)^2} + \tfrac{2}{n} \right] (\Delta f)^2 \leq 0.$$

We choose the following values for δ and m:

$$\delta < \tfrac{1}{2(n-1)} \qquad \text{and} \qquad m = 4(n-1)C.$$

We use the next inequality to simplify the foregoing one

$$\left(\tfrac{1}{n-1} - \delta \right) \left[\tfrac{m}{(C-f)^2} + \tfrac{2}{1+f} \right]^2 - \tfrac{2m}{(C-f)^3} \geq \left(\tfrac{1}{2(n-1)} - \delta \right) \left[\tfrac{m}{(C-f)^2} + \tfrac{2}{1+f} \right]^2,$$

and additionally use the abbreviations:

$$g := \tfrac{m}{(C-f)^2} + \tfrac{2}{1+f}, \quad a := \tfrac{1}{2(n-1)} - \delta > 0, \quad b := \tfrac{4-4(n-1)\delta}{\delta (n-1)^2} + \tfrac{2}{n} > 0.$$

Then we arrive at the inequality:

$$a \cdot g^2\, (f_{,1})^4 - (g \cdot |\Delta f| - nL_1 + (n+2)\, B_{11}) \cdot (f_{,1})^2 - b (\Delta f)^2 \leq 0.$$

The left hand side is a quadratic expression in $(f_{,1})^2$. Consider its zeros, it follows that

$$(f_{,1})^2 \leq \tfrac{1}{a\, g^2} \left[g\, |\Delta f| + | - nL_1 + (n+2)\, B_{11}| + g\sqrt{ab} \cdot |\Delta f| \right].$$

To further estimate this expression we use the assumptions, (4.2.2) and the definition of G; we have the three inequalities

$$g \geq \tfrac{2}{1+f}, \qquad |\Delta f| \leq N^{\sharp}(1+f),$$

$$|-nL_1 + (n+2)\,B_{11}| \leq 2(n+1)\cdot \|B\|_G \leq 2(n+1)N,$$

and insert these inequalities into the foregoing to get an upper bound for $(f_{,1})^2$:

$$(f_{,1})^2 \leq \tfrac{1}{2a}\left(N^{\sharp}(1+\sqrt{ab}) + (n+1)N\right)(1+f)^2 .$$

With our special choice of δ and m and from the definition of ψ we finally get:

$$\psi \leq \tfrac{1}{2a}\left(N^{\sharp}(1+\sqrt{ab}) + (n+1)\,N\right);$$

this inequality holds at x^*, where ψ attains its supremum. Hence, at any point of $S_f(0, C)$, we have

$$\tfrac{\|\mathrm{grad}\, f\|^2}{(1+f)^2} \leq \tfrac{1}{2a}\left(N^{\sharp}(1+\sqrt{ab}) + (n+1)\,N\right)\exp\left\{\tfrac{4(n-1)C}{C-f}\right\}.$$

Let $C \to \infty$, then

$$\tfrac{\|\mathrm{grad}\, f\|}{1+f} \leq \exp\{2(n-1)\}\sqrt{\tfrac{1}{2a}\left(N^{\sharp}(1+\sqrt{ab}) + (n+1)\,N\right)} := Q, \qquad (4.2.14)$$

where Q is a constant. This proves the assertion **(ii)** in the Estimate Lemma. ■

4.2.4 *Topology and the equiaffine Gauß map*

There are several results on geometric properties of the equiaffine Weingarten operator, but sometimes assumptions on it - like that B is positive definite (see e.g. [40], [66], [90]) - seemed to be of more technical character. Thus, for a better geometric understanding, we list some known local properties:
(i) *The condition* $\mathrm{rank}\, S(e) = n$ *is gauge invariant;*
(ii) $im\, \tilde{R} \subset im\, B^{\sharp}$,
(iii) *if* $\mathrm{rank}\, B = n$ *then both Gauß mappings* $Y : M \to V$ *and* $U : M \to V^*$ *are immersions and* B *can be interpreted as their equi-centroaffine "spherical" metric;*
(iv) $(n-1)B = Ric^*$; *this relation generalizes a well known property of the spherical metric in Euclidean geometry, namely that it is a metric of constant sectional curvature (thus, like in the affine case, the spherical metric coincides with the Ricci tensor of the conormal connection modulo a non-zero constant factor).*

Moreover, there are global properties of the Gauß map that are analogues to results of Osserman et al. in the Euclidean theory of minimal surfaces; the following is a typical example, for details see [58], p.233. We will define the notion *affine maximal* in the next chapter, section 5.2.

Theorem. *Let* $x : M \to A^3$ *be a locally strongly convex, affine complete, affine maximal surface. If the Gauß map omits 4 or more points in general position together with their antipodal points, then* $x(M)$ *is an elliptic paraboloid.*

The affine Gauß map and completeness

As before we restrict to locally strongly convex hypersurfaces with Blaschke structure. We recall

a. the Completeness Criterion from above,

b. the subsection "Completeness and Maximum Principles" from section 1.2,

c. the relation for the Ricci tensor from section 2.5.2:

$$R_{ik} = \sum A^m_{il} A^l_{mk} + \tfrac{n-2}{2} B_{ik} + \tfrac{n}{2} L_1 G_{ik}.$$

In this relation, the expression $\sum A^m_{il} A^l_{mk}$ is always non-negative, and the metric G is positive definite. Now assume that the Weingarten form is bounded below, say $B \geq \delta \cdot G$ for some $\delta \in \mathbb{R}$. Then the affine mean curvature is bounded below: $L_1 \geq \delta$. This finally implies that the Ricci tensor is bounded from below. Thus we can state a few additional properties that emphasize some analytical and topological properties.

(i) If $n \geq 2$, if additionally the Weingarten form B is bounded from below and (M, G) is complete and non-compact then one can apply the Maximum Principle of Omori-Yau [104]. Moreover, if B is positive semi-definite and the affine mean curvature L_1 is positively bounded from below then the Ricci curvature is positively bounded from below and then metric completeness implies compactness.

In dimension $n = 2$ we do not need an assumption on B; it is sufficient to assume that L_1 is bounded from below to apply Omori-Yau; moreover, if L_1 positively bounded from below this implies compactness.

(ii) If $\|B\|$ is bounded from above then any Euclidean complete hypersurface is affine complete. In particular, this yields for any global graph over $\mathbb{R}^n$.

(iii) If B satisfies $\delta_1 \cdot G \leq B \leq \delta_2 \cdot G$ for $\delta_1 \in \mathbb{R}$ and $0 < \delta_2 \in \mathbb{R}$ and M is Euclidean complete, then (M, G) is affine complete and one can apply the extended Maximum Principle for weak solutions on (M, G), see [18]. In dimension $n = 2$, in analogy to the statement in (i) above, we can restrict the assumption to L_1.

The affine Gauß map and compactness

The following is another consequence from the above relation between the Ricci tensor *Ric* and the affine spherical metric B:

If (M, G) is complete and B is positively bounded from below then its Ricci tensor is positively bounded from below, thus M is compact (Myers' Theorem). In dimension $n = 2$ it is sufficient to assume that L_1 is positively bounded from below.

In this context, we would like to recall the Gauß conormal equation from section 3.2.1 and the relation $Ric^* = (n-1)B$. In particular we see the following: *If Ric^* is (positively) bounded from below then Ric is (positively) bounded from below.* This shows how curvature properties of ∇^* and the topology of M are related.

The foregoing statement raises the question whether B is always (positive) definite on hyperovaloids; this was stated by Santalo [79]. R. Schneider gave the following counterexample in [80], pp. 84-86. We sketch his construction as it is not printed in a journal.

Counterexample. Let $\mathbb{R}^3 := \{(x^1, x^2, x^3) \mid x^i \in \mathbb{R}\}$ be equipped with a Euclidean inner product $\langle\ ,\ \rangle : \mathbb{R}^3 \times \mathbb{R}^3 \to \mathbb{R}$, thus we can identify $\mathbb{R}^3$ and its dual vector space. In the (x^2, x^3)-plane consider three points defining a triangle, e.g.

$$P_- = (0, -a, 0),\ P_+ = (0, +a, 0),\ P_0 = (0, 0, 10\,a).$$

Rotate this triangle around the x^3-axis; this gives a convex body with a surface C (cut piece of a cone) as boundary; this surface is analytic almost everywhere. We apply an approximation theorem of Minkowski, see [12], p.36:
C can be approximated by a sequence $\{C_i\}_{i\in\mathbb{N}}$ of convex bodies; as boundaries they have analytic ovaloids $\{B_i\}_{i\in\mathbb{N}}$, that means their Euclidean Gauß curvature K_i is positive. Consider the three unit vectors in the (x^2, x^3)-plane in direction to the outside of the given triangle:

$$E_0 := (0, 0, 1), \quad E_1 \perp \text{line}\,(P_-, P_0), \quad E_2 \perp \text{line}\,(P_+, P_0).$$

On each ovaloid $\{B_i\}$ there is exactly one point $P_{i,0},\ P_{i,1},\ P_{i,2}$ with prescribed unit normal $E_0,\ E_1,\ E_2$, respectively. Thus we have three sequences with

$$\lim_i P_{i,0} = P_0, \quad \lim_i P_{i,-} = P_-, \quad \lim_i P_{i,+} = P_+$$

and for the corresponding Euclidean Gauß curvatures:

$$\lim_i K_{i,0} = \infty, \quad \lim_i K_{i,-} = 0, \quad \lim_i K_{i,+} = 0.$$

We recall the section about the affine Gauß map and Euclidean structures. The conormal indicatrices B_i^* of B_i satisfy: $U_i = K_i^{\frac{1}{n+2}} \cdot \mu_i$. Here μ_i is the unit normal of B_i. We arrive at

$$\lim_i \|U_{i,0}\| = 0 \quad \text{and} \quad \lim_i \|U_{i,+}\| = \lim_i \|U_{i,-}\| = \infty.$$

The last relations imply that, for sufficiently large i, the closed conormal indicatrices B_i^* cannot be anymore ovaloids. As already stated, this then yields also for the corresponding normal indicatrices.

The final question in this context is now: what do we know about B on hyperovaloids?

Proposition. *On any hyperovaloid there are open subsets where B is (positive) definite. On such subsets the curvature functions L_r are positive for all $r = 1, ..., n$.*

Proof. We consider local maxima of the affine support function Λ; in such a point we have:

$$\Lambda B = G - Hess\,\Lambda > 0. \qquad \blacksquare$$

Corollary. *If $L_n \neq 0$ on a hyperovaloid then B is positive definite.*

Theorem. *There are no hyperovaloids satisfying $L_1 \leq 0$ on M.*

4.3 Affine Complete Elliptic Affine Hyperspheres

Using (4.1.7), we prove the following theorem of Blaschke ($n = 2$) and Deicke ($n \geq 2$):

Theorem. *Let $x : M \to A^{n+1}$ be a compact affine hypersphere without boundary. Then $x(M)$ is an ellipsoid.*

Proof. We recall (2.4.5) and the fact that, on any ovaloid, there are points such that B is (positive) definite. Since an affine hypersphere satisfies $L_1 = const$, we have $L_1 > 0$. It follows from (4.1.7) that J is a subharmonic function on a compact manifold without boundary. The maximum principle implies $J = const$, and (4.1.7) gives $J = 0$; therefore x is a compact quadric and thus an ellipsoid. ■

Corollary. (i) *Let M be an affine complete elliptic affine hypersphere. Then M is an ellipsoid.*
(ii) *Let M be a Euclidean complete elliptic affine hypersphere. Then M is an ellipsoid.*

Proof. (i) The hypersphere is elliptic, thus $L_1 = const > 0$. Then the Ricci tensor is positively bounded from below, and Myers' Theorem implies that M is compact.
(ii) Apply the foregoing Corollary and the statement of Cheng and Yau in section 4.2.2. ■

4.4 The Theorem of Jörgens-Calabi-Pogorelov

The purpose of this section is to prove a Theorem of Jörgens-Calabi-Pogorelov (see [76] for the result in any dimension). Our proof here is based on [20] and results of Cheng and Yau [25]. Another proof of this Theorem was given by Jost and Yin in [48].

Theorem. (Jörgens-Calabi-Pogorelov). *Let $f : \mathbb{R}^n \to \mathbb{R}$ with $x := (x^1, \cdots, x^n) \mapsto f(x)$ be a strictly convex differentiable function defined for all $x \in \mathbb{R}^n$. If f satisfies the Monge-Ampère equation*

$$\det\left(\frac{\partial^2 f}{\partial x^i \partial x^j}\right) = 1$$

then f must be a quadratic polynomial.

The Theorem of Jörgens-Calabi-Pogorelov concerns the solution of a certain PDE without specific geometric context, but the following proof uses tools from equiaffine hypersurface theory.

Proof. We indicate the steps of the proof.
Step 1. Consider the convex graph M:

$$M := \left\{ \left(x^1, \cdots, x^n, x^{n+1}\right) \mid x^{n+1} = f(x), \;\; x = \left(x^1, \cdots, x^n\right) \in \mathbb{R}^n \right\}.$$

We equip this hypersurface with the Blaschke geometry. We know:

- From the PDE the hypersurface M is a parabolic affine hypersphere with constant affine normal. We aim to prove that the affine sphere is also a quadric, then it must be an elliptic paraboloid and the assertion will be proved.
- By our assumptions M is Euclidean complete; the Cheng-Yau Completeness Criterion states that M is also affine complete.
- For an improper affine sphere, from the integrability conditions we see that the Ricci curvature satisfies $Ric(G) \geq 0$.

Step 2. To prove that M is a quadric we have to show that the Pick invariant vanishes identically. From formula (4.1.7) we can estimate ΔJ from below. We are going to derive a second PDE for J, giving an estimate for ΔJ from above. We denote by r the geodesic distance function (with respect to the Blaschke metric G) from a fixed point p_0. For any $a > 0$, let $\bar{B}_a(p_0, G) = \{p \in M \mid r(p) \leq a\}$ be the closed geodesic ball with radius a around p_0. Define $F : \bar{B}_a(p_0, G) \to \mathbb{R}$ by

$$F(p) := (a^2 - r^2(p))^2 J(p),$$

where J denotes the Pick invariant in the Blaschke geometry (see section 2.4.2). F attains its supremum at some interior point p^* of $\bar{B}_a(p_0, G)$. We may assume that r^2 is a C^2-function in a neighborhood of p^*, and that $J(p^*) > 0$. We choose a local orthonormal frame field of the Blaschke metric. Then, at p^*,

$$F_{,i} = 0, \quad \text{and} \quad \sum F_{,ii} \leq 0. \tag{4.4.1}$$

To derive a second PDE for ΔJ, we calculate both expressions explicitly; as before the norm is defined in terms of the Blaschke metric G:

$$\frac{J_{,i}}{J} - \frac{(2r^2)_{,i}}{a^2 - r^2} = 0, \tag{4.4.2}$$

$$\frac{\Delta J}{J} - \sum \frac{(J_{,i})^2}{J^2} - 2\sum \frac{[(r^2)_{,i}]^2}{(a^2-r^2)^2} - 2\,\frac{\Delta(r^2)}{a^2-r^2} \leq 0. \tag{4.4.3}$$

We insert (4.4.2) into (4.4.3) and get

$$\frac{\Delta J}{J} \leq 24\,\frac{r^2 \|\mathrm{grad}\, r\|^2}{(a^2-r^2)^2} + 4\,\frac{\|\mathrm{grad}\, r\|^2}{a^2-r^2} + 4\,\frac{r\Delta r}{a^2-r^2}\,. \tag{4.4.4}$$

From step 1 recall that (M, G) is a complete Riemann manifold with nonnegative Ricci curvature. Thus the Laplacian Comparison Theorem (see section 1.3) implies the inequality:

$$r\Delta r \leq n - 1. \tag{4.4.5}$$

Now we discuss this differential inequality (4.4.4):
If $p^* = p_0$ then we have $r(p_0, p^*) = 0$.
Otherwise, if $p^* \neq p_0$, from (4.4.4) and the estimate $r\Delta r \leq n-1$, it follows that

$$\frac{\Delta J}{J} \leq 24\,\frac{r^2}{(a^2-r^2)^2} + \frac{4n}{a^2-r^2}\,. \tag{4.4.6}$$

(4.4.6) obviously holds also if $p^* = p_0$.

Step 3. From Step 2 we have two estimates for ΔJ, one from below and one from above. We combine both differential inequalities and get:

$$2(n+1)J \leq 24 \frac{r^2}{(a^2-r^2)^2} + \frac{4n}{a^2-r^2} . \tag{4.4.7}$$

Multiply both sides of (4.4.7) by $(a^2 - r^2)^2$. At p^* we obtain

$$(a^2 - r^2)^2 \, J \leq \frac{2(n+6)}{n+1} \, a^2 . \tag{4.4.8}$$

(4.4.8) holds at p^* where F attains its supremum. Hence, at any interior point of $\bar{B}_a(p_0, G)$, we finally arrive at the upper estimate:

$$J \leq \frac{2(n+6)}{n+1} \cdot \frac{a^2}{(a^2-r^2)^2} = \frac{2(n+6)}{n+1} \cdot \frac{1}{a^2\left(1-\frac{r^2}{a^2}\right)^2} .$$

Step 4. If $a \to \infty$ then $J \to 0$. Hence the Blaschke cubic form satisfies $A \equiv 0$, and that means the improper affine sphere is a quadric. The improper affine sphere has a constant field of affine normals, and therefore we can apply the calculation of A in terms of the graph of the function from section 3.3.4. The cubic form satisfies

$$0 \equiv A_{ijk} = -\tfrac{1}{2} f_{ijk}.$$

Thus f is a quadratic polynomial. This completes the proof of the Theorem of Jörgens-Calabi-Pogorelov in any dimension. ■

4.5 An Extension of the Theorem of Jörgens-Calabi-Pogorelov

To extend the result from the foregoing section, it is natural to study geometric situations where the Monge-Ampère equation in the Theorem of Jörgens-Calabi-Pogorelov appears in a more general form; in particular one will try to find geometric situations where the constant in this PDE is replaced by a suitable function. To make this investigation plausible, we summarize some background from Affine Kähler geometry.

4.5.1 *Affine Kähler Ricci flat equation*

The Theorem of Jörgens-Calabi-Pogorelov was recently extended by L. Caffarelli and Y.Y. Li [16]. We are going to give another extension.

Theorem 4.5.1. [61]. *Let $u : \mathbb{R}^n \to \mathbb{R}$, $(\xi_1, ..., \xi_n) \mapsto u(\xi_1, ..., \xi_n)$, be a strictly convex C^∞-function. If u satisfies the PDE of Monge-Ampère type*

$$\det\left(\frac{\partial^2 u}{\partial \xi_i \partial \xi_j}\right) = \exp\left\{-\sum c_i \frac{\partial u}{\partial \xi_i} - c_0\right\}, \tag{4.5.1}$$

where c_0, c_1,...,c_n are real constants, then u must be a quadratic polynomial.

A more precise statement of the assertion in the theorem says that there exists a solution of the PDE (4.5.1) defined on $\mathbb{R}^n$ only if $c_1 = ... = c_n = 0$.

Our proof will show that we can state a stronger version of the foregoing theorem as follows.

The Extended Theorem for domains. *Let $u(\xi_1, ..., \xi_n)$ be a strictly convex C^∞-function defined on a convex domain $\Omega \subset \mathbb{R}^n$. If $u(\xi)$ satisfies the PDE (4.5.1), and if $u(\xi) \to \infty$ for $\xi \to \partial\Omega$, then u must be a quadratic polynomial.*

Remark. While we proved the Theorem of Jörgens-Calabi-Pogorelov with geometric tools from unimodular hypersurface theory, now we will use Calabi's relative normalization as tool for the proof. This way we demonstrate the appropriate *geometric modelling* for solving certain classes of PDEs. As already stated, before giving the proof, we consider another geometric background of the PDE (4.5.1).

A background from Affine Kähler geometry. The Monge-Ampère equation (4.5.1) has another geometric background in affine Kähler geometry. Consider the Legendre transformation of u from section 1.4; in terms of $x^1, ..., x^n$ and $f(x^1, ..., x^n)$, the PDE (4.5.1) can be written as

$$\det\left(\tfrac{\partial^2 f}{\partial x^i \partial x^j}\right) = \exp\left\{\sum c_i x^i + c_0\right\}, \tag{4.5.2}$$

or equivalently

$$\tfrac{\partial^2}{\partial x^i \partial x^j}\left(\ln\det\left(\tfrac{\partial^2 f}{\partial x^k \partial x^l}\right)\right) = 0. \tag{4.5.3}$$

Let M be a graph defined by the function $x^{n+1} = f(x^1, ..., x^n)$, namely,

$$M := \{\, (x, f(x)) \mid x \in \mathbb{R}^n\}.$$

Following [19] and [76], we consider the Calabi metric from section 3.3.4:

$$\mathfrak{H} := \sum f_{ij}\, dx^i dx^j. \tag{4.5.4}$$

We note that any affine Kähler manifold can be considered as a totally real submanifold of a complex Kähler manifold in the following way. For each coordinate chart $(x^1, x^2, \ldots, x^n)$, we can consider a tube over the coordinate neighborhood with a complex coordinate system $(x^1 + \sqrt{-1}y^1, x^2 + \sqrt{-1}y^2, \ldots, x^n + \sqrt{-1}y^n)$. The affine coordinate transformations naturally piece together these tubes to form a complex manifold. The Hessian metric $\mathfrak{H}$ can be naturally extended to a Kähler metric of the complex manifold. The Ricci tensor $Ric(\mathfrak{H})$ and the normed scalar curvature $\kappa(\mathfrak{H})$ of this Kähler metric satisfy the relations

$$R(\mathfrak{H})_{ij} = -\tfrac{\partial^2}{\partial x^i \partial x^j}\left(\ln\det\left(f_{kl}\right)\right),$$
$$n(n-1)\kappa(\mathfrak{H}) = -\tfrac{1}{2}\sum_{i,j} f^{ij}\, \tfrac{\partial^2\left(\ln[\det(f_{kl})]\right)}{\partial x^i \partial x^j}.$$

We call $Ric(\mathfrak{H})$ and $\kappa(\mathfrak{H})$ the *affine Kähler Ricci curvature* and the *affine Kähler scalar curvature* of the affine Kähler metric, resp. We state:

Regarding equation (4.5.3), it follows that the affine Kähler metric $\mathfrak{H}$ is Ricci flat.

Remark. We point out that we do not have global uniqueness for solutions of the equation (4.5.2) on $\mathbb{R}^n$. For example, the functions $f,\ f^\sharp : \mathbb{R}^n \to \mathbb{R}$, given by

$$f(x^1, ..., x^n) := \sum_{i=1}^{n} (x^i)^2 \quad \text{and} \quad f^\sharp(x^1, ..., x^n) := \exp\{x^1\} + \sum_{i=2}^{n} (x^i)^2,$$

both satisfy (4.5.2).

4.5.2 *Tools from relative geometry*

As indicated in the beginning of Chapter 4, to solve the Monge-Ampère equation in Theorem 4.5.1, we use tools from a different relative hypersurface geometry in Chapter 3. Now we will apply Calabi's geometry.
Let f be a strictly convex C^∞ function defined on a domain $\Omega \subset \mathbb{R}^n$, and consider the graph hypersurface

$$M := \{(x, f(x)) \mid x^{n+1} = f(x^1, ..., x^n), \quad (x^1, ..., x^n) \in \Omega\}.$$

For M we choose the canonical relative normalization of Calabi, given by $Y = (0, 0, ..., 1)$. In section 3.3.4 we listed basic geometric invariants up to the calculation of the Laplacian. We apply this to calculate the Laplacian for the functions u and its Legendre transform, the graph function f; for this we recall the notation from section 1.4; we recall the involution in the choice of coordinate systems.

$$\rho := \left[\det\left(\tfrac{\partial^2 u}{\partial \xi_i\, \partial \xi_j}\right)\right]^{\frac{1}{n+2}} = \left[\det\left(\tfrac{\partial^2 f}{\partial x^i\, \partial x^j}\right)\right]^{\frac{-1}{n+2}}$$

then in terms of the coordinates x:

$$\Delta f = n + \tfrac{n+2}{2\rho}\, \mathfrak{H}(\operatorname{grad}\rho, \operatorname{grad} f), \tag{4.5.5}$$

and in terms of the coordinates ξ:

$$\Delta u = n - \tfrac{n+2}{2\rho}\, \mathfrak{H}(\operatorname{grad}\rho, \operatorname{grad} u). \tag{4.5.6}$$

4.5.3 *Calculation of $\Delta\Phi$ in terms of the Calabi metric*

In the next step, we recall the definition of the second auxiliary function Φ from section 1.4, given in terms of u and f, resp., and thus for both possible choices of coordinates.

The geometric meaning of Φ.

As we are using Calabi's relative normalization, we would like to point out the geometric meaning of Φ:

$$\Phi = \tfrac{4n^2}{(n+2)^2} \|T\|^2.$$

As stated in section 3.3.2, the Tchebychev vector field T of a relative hypersurface (x, U, Y) vanishes identically if and only if the normalization is the Blaschke normalization; as we apply Calabi's normalization, this means that, for $\Phi = 0$, Calabi's and Blaschke's normalization coincide; that means that the hypersurface is an improper affine sphere. From the definition of ρ, this finally will give:

$$\det(u_{ij}) = const$$

everywhere on M. This is the original Monge-Ampère equation for improper affine hyperspheres. Thus, from the geometric meaning, we see that our aim is to prove $\Phi = 0$.

Reformulation of the PDE.

We consider the extended PDE (4.5.1) and reformulate it in terms of its Legendre transform function $f = f(x)$: (4.5.2) or equivalently (4.5.3). From this we reformulate the original PDE (4.5.1) in terms of the auxiliary function ρ as follows:

$$0 = \frac{\partial^2}{\partial x^i \partial x^j}\left(\ln\det\left(f_{kl}\right)\right) = -(n+2)\left(\frac{\rho_{ij}}{\rho} - \frac{\rho_i}{\rho}\cdot\frac{\rho_j}{\rho}\right); \tag{4.5.7}$$

finally we get:

$$\Delta\,\rho = \frac{n+4}{2}\,\frac{\|\mathrm{grad}\,\rho\|^2}{\rho}. \tag{4.5.8}$$

In the following, let $f(x^1, ..., x^n)$ be a smooth, strictly convex solution of the PDE

$$\Delta\rho = -\beta\,\frac{\|\mathrm{grad}\,\rho\|^2}{\rho}, \tag{4.5.9}$$

where ρ is defined above, and $\beta \in \mathbb{R}$ is a constant. Recall section 1.4 and the fact that the PDE (4.5.1) can be expressed in both coordinate systems, in terms of (x) as well as in terms of (ξ). For $f(x^1, ..., x^n)$, we shall derive an estimate for $\Delta\Phi$.

Proposition 4.5.2. *Let $f(x^1, ..., x^n)$ be a strictly convex C^∞ function satisfying the PDE (4.5.9). In terms of the Calabi metric, the Laplacian of Φ satisfies the following inequality:*

$$\begin{aligned}\Delta\Phi \geq \frac{2\delta}{\rho^2}\sum(\rho_{,ij})^2 + \frac{n(1-\delta)}{2(n-1)}\frac{\|\mathrm{grad}\Phi\|^2}{\Phi} - \left(\frac{2\beta(n-2+\delta)}{n-1} + \frac{2n+2n\delta-4}{n-1}\right)\mathfrak{H}(\mathrm{grad}\Phi, \mathrm{grad}\ln\rho) \\ + \left(\frac{(2\beta^2+4\beta)(1-\delta)+2-2n\delta}{n-1} - \frac{(n+2)^2(n-1)}{8n}\right)\Phi^2\end{aligned}$$

for any $0 \leq \delta < 1$.

Proof. At $p \in M$ we choose a local orthonormal frame field of the metric $\mathfrak{H}$. Then

$$\Phi = \sum\frac{(\rho,_j)^2}{\rho^2}, \qquad \Phi,_i = 2\sum\frac{\rho,_j\cdot\rho,_{ji}}{\rho^2} - 2\rho,_i\sum\frac{(\rho,_j)^2}{\rho^3},$$

and

$$\Delta\Phi = 2\sum\frac{(\rho,_{ji})^2}{\rho^2} + 2\sum\frac{\rho,_j\cdot\rho,_{jii}}{\rho^2} - 8\sum\frac{\rho,_j\cdot\rho,_i\cdot\rho,_{ji}}{\rho^3} + (6+2\beta)\Phi^2,$$

where we used (4.5.9).

We discuss two cases, namely $\Phi(p) = 0$ and $\Phi(p) \neq 0$.

1. Assume $\Phi(p) = 0$, then Φ takes a minimum at p and thus $\mathrm{grad}\,\rho(p) = 0$: then, at p,

$$\Delta\Phi \geq 2\sum\frac{(\rho,_{ij})^2}{\rho^2}.$$

2. Assume $\Phi(p) \neq 0$. Around p we choose a local orthonormal frame field of the metric $\mathfrak{H}$ such that $\rho_{,1}(p) = \|\operatorname{grad}\rho\|(p) > 0$, $\rho_{,i}(p) = 0$ for all $i > 1$. Then

$$\Delta\Phi = 2(1-\delta+\delta)\sum \tfrac{(\rho_{,ij})^2}{\rho^2} + 2\sum \tfrac{\rho_{,j}\rho_{,jii}}{\rho^2} - 8\tfrac{(\rho_{,1})^2\rho_{,11}}{\rho^3} + (6+2\beta)\Phi^2 \tag{4.5.10}$$

for any $0 \leq \delta < 1$. Applying the inequality of Schwarz and (4.5.9) we get

$$\begin{aligned} 2\sum(\rho_{,ij})^2 &\geq 2(\rho_{,11})^2 + 4\sum_{i>1}(\rho_{,1i})^2 + \tfrac{2}{n-1}(\Delta\rho - \rho_{,11})^2 \\ &= \tfrac{2n}{n-1}(\rho_{,11})^2 + 4\sum_{i>1}(\rho_{,1i})^2 + \tfrac{4\beta}{n-1}\tfrac{(\rho_{,1})^2\rho_{,11}}{\rho} + \tfrac{2\beta^2}{n-1}\tfrac{(\rho_{,1})^4}{\rho^2}. \end{aligned} \tag{4.5.11}$$

An application of the Ricci identities and (4.5.9) shows that

$$\tfrac{2}{\rho^2}\sum \rho_{,j}\rho_{,jii} = -4\beta\,\tfrac{(\rho_{,1})^2\rho_{,11}}{\rho^3} + 2\beta\Phi^2 + 2R_{11}\tfrac{(\rho_{,1})^2}{\rho^2}. \tag{4.5.12}$$

We insert (4.5.11) and (4.5.12) into (4.5.10) and obtain

$$\begin{aligned} \Delta\Phi \geq{}& \tfrac{2\delta}{\rho^2}\sum(\rho_{,ij})^2 + \tfrac{2n(1-\delta)}{n-1}\tfrac{(\rho_{,11})^2}{\rho^2} + \tfrac{4(1-\delta)}{\rho^2}\sum_{i>1}(\rho_{,1i})^2 + 2R_{11}\tfrac{(\rho_{,1})^2}{\rho^2} \\ &- \left(\tfrac{4\beta(n-2+\delta)}{n-1} + 8\right)\tfrac{(\rho_{,1})^2\rho_{,11}}{\rho^3} + \left(\tfrac{2\beta^2(1-\delta)}{n-1} + 6 + 4\beta\right)\Phi^2. \end{aligned} \tag{4.5.13}$$

Note that

$$\Phi_{,i} = 2\,\tfrac{\rho_{,1}\rho_{,1i}}{\rho^2} - 2\,\tfrac{\rho_{,i}(\rho_{,1})^2}{\rho^3}.$$

Hence

$$\sum \tfrac{(\Phi_{,i})^2}{\Phi} = 4\sum \tfrac{(\rho_{,1i})^2}{\rho^2} - 8\,\tfrac{(\rho_{,1})^2\rho_{,11}}{\rho^3} + 4\,\tfrac{(\rho_{,1})^4}{\rho^4}. \tag{4.5.14}$$

Then (4.5.13) and (4.5.14) together give

$$\begin{aligned} \Delta\Phi \geq{}& \tfrac{n(1-\delta)}{2(n-1)}\sum \tfrac{(\Phi_{,i})^2}{\Phi} - \left(\tfrac{2\beta(n-2+\delta)}{n-1} + 4 - \tfrac{2n(1-\delta)}{n-1}\right)\Phi_{,1}\tfrac{\rho_{,1}}{\rho} \\ &+ \tfrac{2\delta}{\rho^2}\sum(\rho_{,ij})^2 + 2R_{11}\tfrac{(\rho_{,1})^2}{\rho^2} + \tfrac{(2\beta^2+4\beta)(1-\delta)+2-2n\delta}{n-1}\Phi^2, \end{aligned} \tag{4.5.15}$$

where we use the relation

$$\tfrac{(\rho_{,1})^2\rho_{,11}}{\rho^3} = \tfrac{1}{2}\Phi_{,1}\tfrac{\rho_{,1}}{\rho} + \Phi^2.$$

Using the same method as in deriving (4.5.11), we get

$$\begin{aligned} \sum(A_{ml1})^2 &\geq (A_{111})^2 + 2\sum_{i>1}(A_{i11})^2 + \tfrac{1}{n-1}\left(\sum A_{ii1} - A_{111}\right)^2 \\ &\geq \tfrac{n}{n-1}\sum(A_{i11})^2 - \tfrac{2}{n-1}A_{111}\sum A_{ii1} + \tfrac{1}{n-1}\left(\sum A_{ii1}\right)^2. \end{aligned}$$

Note that $\sum A_{ii1} = \frac{n+2}{2}\frac{\rho_1}{\rho}$. Therefore

$$\begin{aligned} 2R_{11}\tfrac{(\rho_{,1})^2}{\rho^2} &= 2\sum(A_{ml1})^2\tfrac{(\rho_{,1})^2}{\rho^2} - (n+2)A_{111}\tfrac{(\rho_{,1})^3}{\rho^3} \\ &\geq \tfrac{2n}{n-1}\sum(A_{i11})^2\tfrac{(\rho_{,1})^2}{\rho^2} - \tfrac{(n+1)(n+2)}{n-1}A_{111}\tfrac{(\rho_{,1})^3}{\rho^3} + \tfrac{(n+2)^2}{2(n-1)}\Phi^2. \end{aligned} \tag{4.5.16}$$

The inequality of Schwarz gives

$$2R_{11}(p)\frac{(\rho_{,1})^2}{\rho^2} \geq -\frac{(n+2)^2(n-1)}{8n}\Phi^2. \qquad (4.5.17)$$

A combination of (4.5.15) and (4.5.17) yields Proposition 4.5.2. ■

Proposition 4.5.3. *Let f be a strictly convex C^∞ function satisfying the PDE (4.5.2). Then the Laplacian of Φ satisfies the following inequality:*

$$\Delta\Phi \geq \frac{n}{n-1}\frac{\|\mathrm{grad}\,\Phi\|^2}{\Phi} + \frac{n^2-3n-10}{2(n-1)}\mathfrak{H}(\mathrm{grad}\,\Phi,\ \mathrm{grad}\ln\rho) + \frac{(n+2)^2}{n-1}\Phi^2.$$

Proof. We reformulated the PDE (4.5.2), and arrived at (4.5.8). In (4.5.15) we choose $\beta = -\frac{n+4}{2}$ and $\delta = 0$ and get

$$\Delta\Phi \geq \frac{n}{2(n-1)}\sum\frac{(\Phi_{,i})^2}{\Phi} + \frac{n^2-4}{n-1}\frac{\Phi_{,1}\cdot\rho_{,1}}{\rho} + 2R_{11}\frac{(\rho_{,1})^2}{\rho^2} + \frac{(n+2)^2}{2(n-1)}\Phi^2. \qquad (4.5.18)$$

In the following we calculate the term $2R_{11}\frac{(\rho_{,1})^2}{\rho^2}$. Choose coordinates $x^1, ..., x^n$ around p such that $f_{ij}(p) = \delta_{ij}$ and $\frac{\partial\rho}{\partial x^1} = \|\mathrm{grad}\,\rho\|(p) > 0$, $\frac{\partial\rho}{\partial x^i}(p) = 0$ for all $i > 1$. From (4.5.7) we easily obtain

$$\rho_{,ij} = \rho_{ij} - \Gamma_{ij}^k\rho_k = \rho_{ij} + A_{ij1}\rho_{,1} = \frac{\rho_{,i}\rho_{,j}}{\rho} + A_{ij1}\rho_{,1}. \qquad (4.5.19)$$

We insert (4.5.19) into the foregoing expression for $\Phi_{,i}$:

$$\Phi_{,i} = 2\,\frac{\rho_{,1}\rho_{,1i}}{\rho^2} - 2\,\frac{\rho_{,i}(\rho_{,1})^2}{\rho^3} = 2\,A_{i11}\frac{(\rho_{,1})^2}{\rho^2}.$$

It follows that

$$\frac{\sum(\Phi_{,i})^2}{\Phi} = 4\sum(A_{i11})^2\frac{(\rho_{,1})^2}{\rho^2}, \qquad \sum\Phi_{,i}\frac{\rho_{,i}}{\rho} = 2A_{111}\frac{(\rho_{,1})^3}{\rho^3}. \qquad (4.5.20)$$

Therefore, by (4.5.16) and (4.5.20), we obtain

$$2R_{11}\frac{(\rho_{,1})^2}{\rho^2} \geq \frac{n}{2(n-1)}\frac{\sum(\Phi_{,i})^2}{\Phi} - \frac{(n+1)(n+2)}{2(n-1)}\sum\Phi_{,i}\frac{\rho_{,i}}{\rho} + \frac{(n+2)^2}{2(n-1)}\Phi^2. \qquad (4.5.21)$$

We insert (4.5.21) into (4.5.18), this finishes the proof of Proposition 4.5.3. ■

4.5.4 *Extension of the Theorem of Jörgens-Calabi-Pogorelov - proof for $n \leq 4$*

In dimension $n \leq 4$ the proof of Theorem 4.5.1 is relatively simple. First we consider this case. We use the Calabi metric; recall section 3.3.4 and the definitions of ρ and Φ. As sketched above we aim to show that $\Phi = 0$ on M everywhere.

Step 1. Subtracting a linear function we may assume that

$$u(0) = 0, \quad \mathrm{grad}\,u(0) = 0, \quad u(\xi) \geq 0, \qquad \forall\xi \in \mathbb{R}^n.$$

Then, for any constant $C > 0$, recall the notation for the level set from section 1.1.4

$$S_u(0, C) := \{\xi \mid u(\xi) < C\}.$$

Step 2. In the next two steps, step 2 and step 3, we derive two differential inequalities for $\Delta\Phi$. To derive an upper bound for $\Delta\Phi$, we use an auxiliary function F on

a level set (step 2). A lower bound for $\Delta\Phi$ in terms of ξ-coordinates follows from Proposition 4.5.3, where we reformulated the PDE (4.5.1) in terms of Φ; here we point out that the function Φ is a geometric invariant that does not depend on the choice of coordinates. Namely, for a function $u = u(\xi)$ satisfying the PDE (4.5.1), we can reformulate Proposition 4.5.3 in terms of the ξ-coordinates.
To derive a differential inequality for $\Delta\Phi$, consider the function

$$F(\xi) := \exp\left\{-\tfrac{m}{C-u}\right\}\Phi,$$

defined on $S_u(0, C)$, where m is a positive constant to be determined later. Clearly, F attains its supremum at some interior point p^*. Then, at p^*,

$$\begin{aligned} &\tfrac{\Phi_{,i}}{\Phi} - \gamma\, u_{,i} = 0, \\ &\tfrac{\Delta\Phi}{\Phi} - \tfrac{\sum(\Phi_{,i})^2}{\Phi^2} - \gamma^{'} \sum(u_{,i})^2 - \gamma\,\Delta u \le 0, \end{aligned} \tag{4.5.22}$$

where the norm is taken with respect to the Calabi metric; we fix the notation

$$\gamma := \tfrac{m}{(C-u)^2} \quad \text{and} \quad \gamma^{'} := 2\tfrac{m}{(C-u)^3}\,.$$

Note that (4.5.22) gives an upper bound for $\Delta\Phi$.
Step 3. Proposition 4.5.3 gives a lower bound for $\Delta\Phi$; we combine both inequalities and insert Proposition 4.5.3 and formula (4.5.6) into (4.5.22); we get

$$\tfrac{(n+2)^2}{n-1}\,\Phi + \left(\tfrac{1}{n-1}\gamma^2 - \gamma^{'}\right)\sum(u_{,i})^2 - n\gamma + \tfrac{(n+2)(n-3)}{n-1}\,\gamma\,\tfrac{\sum \rho_{,i}\,u_{,i}}{\rho} \le 0.$$

We apply the inequality of Schwarz:

$$\tfrac{(n+2)(n-3)}{n-1}\,\gamma\,\tfrac{\sum \rho_{,i}\,u_{,i}}{\rho} \le \tfrac{1}{2(n-1)}\,\gamma^2 \sum(u_{,i})^2 + \tfrac{(n+2)^2(n-3)^2}{2(n-1)}\,\Phi.$$

Therefore

$$\tfrac{(n+2)^2(2-(n-3)^2)}{2(n-1)}\,\Phi + \left(\tfrac{1}{2(n-1)}\gamma^2 - \gamma^{'}\right)\sum(u_{,i})^2 - n\gamma \le 0.$$

Now, if $n \le 4$, this implies an upper bound for Φ in terms of ξ-coordinates:

$$\tfrac{(n+2)^2}{2(n-1)}\,\Phi + \left(\tfrac{1}{2(n-1)}\,\gamma^2 - \gamma^{'}\right)\sum(u_{,i})^2 - n\gamma \le 0.$$

We choose $m = 8(n-1)C$, then the factor satisfies $\tfrac{1}{2(n-1)}\gamma^2 - \gamma^{'} \ge 0$. Finally at p^*:

$$\exp\left\{-\tfrac{8(n-1)C}{C-u}\right\}\Phi \le \tfrac{b}{C}, \tag{4.5.23}$$

where b is a positive constant depending only on the dimension n. In the calculation of (4.5.23) and later we often use the fact that

$$\exp\left\{-\tfrac{m}{C-u}\right\}\tfrac{m^2}{(C-u)^2}$$

has a universal upper bound. Since the function F attains its supremum at p^*, (4.5.23) holds everywhere in $S_u(0, C)$. For any fixed point p, we let $C \to \infty$, then $\Phi(p) = 0$. Therefore $\Phi = 0$ everywhere on M. This finishes the proof in dimension $n \le 4$. ∎

4.5.5 *Comparison of two geometric proofs*

We would like to comment on the proofs of two Theorems, namely that of Jörgens-Calabi-Pogorelov in section 4.4 and its extension for $n \leq 4$ in the foregoing section; see also [89]. While both proofs use geometric modelling techniques and thus are appropriate to demonstrate the geometric ideas, the proof for $n > 4$, in the sections following below, is quite technical and complicated and thus not ideal to survey and comment on geometric modelling techniques.

The Theorem of Jörgens-Calabi-Pogorelov. Our proof of the Theorem of Jörgens-Calabi-Pogorelov starts with the geometric interpretation of the given Monge-Ampère equation (4.1.2), using the local characterization of a locally strongly convex improper affine hypersphere in terms of a PDE, as described in sections 4.1.1 - 4.1.2. We aim to show that the improper affine hypersphere is also a quadric, then it must be an elliptic paraboloid, and we arrive at the polynomial solution.
For our geometric modelling we use Blaschke's geometry. A locally strongly convex quadric can be characterized by the vanishing of its Pick invariant J. Therefore we aim to estimate J from above.
In a first step, we use the known fact that the Euclidean completeness implies the affine completeness (section 4.2.2); we recall that the Ricci curvature is non-negative. In the second step, we use two differential inequalities for the Pick invariant, namely:
(i) For ΔJ, we have the relation in section 4.1.3 as a first differential inequality, giving a lower estimate for ΔJ.
(ii) To derive a second differential inequality for ΔJ in Step 2 of the proof, we define the auxiliary function

$$F(p) := (a^2 - r^2(p_0, p))^2 \cdot J(p)$$

on a geodesic ball $B_a(p_0, G)$ and apply the Laplacian Comparison Theorem; this gives an upper estimate for ΔJ.
The combination of both inequalities gives an upper estimate for $J(p)$ at an arbitrary point $p \in B_a(p_0, G)$:

$$J(p) \leq \tfrac{2(n+6)}{n+1} \cdot \tfrac{a^2}{(a^2-r^2)^2} = \tfrac{2(n+6)}{n+1} \cdot \frac{1}{a^2\left(1-\frac{r^2}{a^2}\right)^2}.$$

In the final step, for $a \to \infty$ (completeness of the Blaschke metric), we get the assertion.

The extended Theorem for $n \leq 4$. For the given PDE (4.5.1) we consider the Legendre transform function f of u and the locally strongly convex graph hypersurface defined by f. Recall that the Legendre transformation is a useful tool when studying Monge-Ampère equations.
Now we use Calabi's graph normalization of f for our geometric modelling (section 3.3.4). We aim to show that the Tchebychev vector field T in this geometry vanishes identically, then the normalization must coincide with Blaschke's normalization (section 3.3.2), and in this geometry a constant affine normal characterizes

an improper affine hypersphere (section 4.4.1). This way we want to reduce the problem to the foregoing Theorem.
Again we derive two differential inequalities, now using the auxiliary function

$$\Phi := \tfrac{4n^2}{(n+2)^2} \cdot \|T\|^2,$$

where the norm is taken with respect to the Calabi metric, and aim to prove $\Phi = 0$. For this we are going to estimate Φ, combining the following steps (i) and (ii):
(i) The first differential inequality for $\Delta\Phi$ directly comes from Proposition 4.5.3, giving a lower estimate for $\Delta\Phi$, where we use the Legendre transformation of f.
(ii) The second differential inequality, giving an upper estimate for $\Delta\Phi$, comes from a derivation of the auxiliary function

$$F(\xi) := \exp\left\{-\tfrac{m}{C-u(\xi)}\right\}\Phi(\xi)$$

on a level set:

$$S_u(0, C) := \{\xi \in \mathbb{R}^n \mid u(\xi) < C\}.$$

We use the involutionary character of the Legendre transformation and combine both differential inequalities; this leads to an upper bound at interior points of the level set:

$$\exp\left\{-\tfrac{8(n-1)C}{C-u}\right\}\Phi \leq \tfrac{b}{C}.$$

We apply the Euclidean completeness; $C \to \infty$ then gives $\Phi = 0$ and thus $T = 0$ on M.

The comparison of both proofs clearly shows the importance of the geometric modelling, the analogies in deriving two different differential inequalities on typical domains (related to the completeness conditions considered), and the final limiting procedure.

4.5.6 *Technical tools for the proof in dimension* $n \geq 5$

In dimension $n > 4$ the proof of the extension of the Theorem of Jörgens-Calabi-Pogorelov is much more difficult than for $n \leq 4$. We will prove several estimates that we will need for the proof of Theorem 4.5.1. As pointed out in the beginning of section 4.5.2, for a given strictly convex function f we consider its Legendre transform function u and also the graph hypersurface, defined by f, together with the canonical normalization introduced by Calabi (sections 3.3.4 and 4.5.2).
Notational agreement. In this section a pair of functions, denoted by u and f, is always a pair of Legendre transformation functions defined on corresponding Legendre transform domains. As before we use

$$\rho := \left[\det\left(\tfrac{\partial^2 u}{\partial\xi_i\,\partial\xi_j}\right)\right]^{\frac{1}{n+2}} = \left[\det\left(\tfrac{\partial^2 f}{\partial x^i\,\partial x^j}\right)\right]^{\frac{-1}{n+2}}$$

and

$$\Phi := \tfrac{\|\mathrm{grad}\,\rho\|^2}{\rho^2}.$$

From section 1.4 recall that we can express such terms in both coordinate systems, denoted by x and ξ.
If $k \in \mathbb{N}$ and we have pairs of functions $(u^{(k)}, f^{(k)})$, an obvious analogous notation yields.

I. A gradient estimate for $\sum(\frac{\partial u}{\partial \xi_i})^2$.

Consider $\mathbb{R}^n$ with the canonical Euclidean structure and a bounded convex domain $\Omega \subset \mathbb{R}^n$. From [37], p.27, it is known that there exists a unique ellipsoid E, which attains the minimal volume among all ellipsoids that contain Ω and that are centered at the center of mass of Ω, such that

$$n^{-\frac{3}{2}}E \subset \Omega \subset E,$$

where $n^{-\frac{3}{2}}E$ means the $(n^{-\frac{3}{2}})$-dilation of E with respect to its center. Let $\mathfrak{T}$ be an affine transformation such that $\mathfrak{T}(E) = B_1(0)$, the Euclidean ball with radius 1 around $0 \in \mathbb{R}^n$. Put $\tilde{\Omega} := \mathfrak{T}(\Omega)$. Then

$$B_{n^{-\frac{3}{2}}}(0) \subset \tilde{\Omega} \subset B_1(0). \tag{4.5.24}$$

A convex domain Ω is called *normalized* if $\mathfrak{T}$ is the identity mapping, and the minimal ellipsoid is the unit ball with center of mass at 0.

Lemma 4.5.4. *Let $\Omega_k \subset \mathbb{R}^n$ be a sequence of smooth normalized convex domains, and $u^{(k)}$ be a sequence of strictly convex smooth functions defined on Ω_k. Assume that*

$$\inf_{\Omega_k} u^{(k)} = u^{(k)}(q_k) = 0, \qquad u^{(k)} = C > 0 \quad \text{on} \quad \partial\Omega_k.$$

Then there are constants $d > C$, $b > 0$, independent of k, such that

$$\frac{\sum_i \left(\frac{\partial u^{(k)}}{\partial \xi_i}\right)^2}{(d+f^{(k)})^2} \leq b, \quad k = 1, 2, ... \quad \text{on} \quad \bar{\Omega}_k,$$

where $f^{(k)}$ is the Legendre transformation function of $u^{(k)}$ relative to 0.

Proof. We may assume (if necessary we consider a subsequence) that the sequence of domains Ω_k converges to a convex domain Ω (for more details see Lemma 5.3.1 in [37]) and the sequence of functions $u^{(k)}$ locally uniformly converges to a convex function u^∞ in Ω. As $\text{dist}(0, \partial\Omega_k) \geq \frac{1}{2}n^{-\frac{3}{2}}$, we have the uniform estimate

$$\sum_{i=1}^{n} \left(\frac{\partial u^{(k)}}{\partial \xi_i}\right)^2 (0) \leq 4n^3C^2. \tag{4.5.25}$$

For any k, define

$$\tilde{u}^{(k)} := u^{(k)} - \sum_{i=1}^{n} \frac{\partial u^{(k)}}{\partial \xi_i}(0)\xi_i - u^{(k)}(0). \tag{4.5.26}$$

Then

$$\tilde{u}^{(k)}(0) = 0, \qquad \tilde{u}^{(k)}(\xi) \geq 0, \qquad \tilde{u}^{(k)}|_{\partial\Omega_k} \leq C_0,$$

where C_0 is a constant depending only on n and C. As $B_{n^{-\frac{3}{2}}}(0) \subset \Omega_k$, we have

$$\frac{\|\operatorname{grad} \tilde{u}^{(k)}\|^2}{(1+\tilde{f}^{(k)})^2} \leq \|\operatorname{grad} \tilde{u}^{(k)}\|^2 \leq \frac{C_0^2}{(d_{n,k})^2} \leq 4n^3 C_0^2$$

on the ball $B_{2^{-1}n^{-\frac{3}{2}}}(0)$, where $\tilde{f}^{(k)}$ is the Legendre transformation of $\tilde{u}^{(k)}$ relative to 0, and $d_{n,k} := \operatorname{dist}\left(B_{2^{-1}n^{-\frac{3}{2}}}(0), \partial\Omega_k\right)$. For any $p \in \bar{\Omega}_k \backslash B_{2^{-1}n^{-\frac{3}{2}}}(0)$ we may assume, by an orthonormal transformation, that p has coordinates $p = (\xi_1, 0, \ldots, 0)$ with $\xi_1 > 0$. Then, at p,

$$C_0 + \tilde{f}^{(k)} \geq \tilde{u}^{(k)} + \tilde{f}^{(k)} = \tfrac{\partial \tilde{u}^{(k)}}{\partial \xi_1} \cdot \xi_1.$$

It follows that

$$\frac{\left(\frac{\partial \tilde{u}^{(k)}}{\partial \xi_1}\right)^2}{(C_0 + \tilde{f}^{(k)})^2} < \frac{1}{\xi_1^2} < 4n^3.$$

Therefore there exist constants $\tilde{d} > 1$, $\tilde{b} > 0$, depending only on n and C, such that

$$\frac{(\frac{\partial \tilde{u}^{(k)}}{\partial r})^2}{(\tilde{d} + \tilde{f}^{(k)})^2} < \tilde{b},$$

where $\frac{\partial}{\partial r}$ denotes the radial derivative. From (4.5.25) and (4.5.26) we get

$$\left(\tfrac{\partial u^{(k)}}{\partial r}\right)^2 \leq 2\left(\tfrac{\partial \tilde{u}^{(k)}}{\partial r}\right)^2 + 8n^3 C^2.$$

Note that

$$\frac{\partial \tilde{u}^{(k)}}{\partial \xi_i} = \frac{\partial u^{(k)}}{\partial \xi_i} - \frac{\partial u^{(k)}}{\partial \xi_i}(0), \qquad \tilde{f}^{(k)} = f^{(k)} + u^{(k)}(0). \tag{4.5.27}$$

Then

$$\frac{\left(\frac{\partial u^{(k)}}{\partial r}\right)^2}{(d' + f^{(k)})^2} < b'$$

for some constants $d' > 1$ and $b' > 0$, independent of k. Note that

$$\|\operatorname{grad} u^{(k)}(p)\| = \tfrac{1}{\cos \alpha_k} \cdot \left\|\tfrac{\partial u^{(k)}}{\partial r}(p)\right\|,$$

where α_k is the angle between the vectors $\operatorname{grad} u^{(k)}(p)$ and $\frac{\partial u^{(k)}}{\partial r}(p)$. Since $u^{(k)} = C$ on $\partial\Omega_k$, the vector $\operatorname{grad} u^{(k)}(p)$ is perpendicular to the boundary of the domain Ω_k at any $p \in \partial\Omega_k$. As Ω is convex and $0 \in \Omega$, it follows that the sequence $\frac{1}{\cos \alpha_k}$ has a uniform upper bound. This proves Lemma 4.5.4. ■

Remark. From the second equation in (4.5.27) we know that

$$f^{(k)} \geq -C, \qquad \text{on} \quad \Omega_k.$$

Thus we may choose d in Lemma 4.5.4 such that the following holds for any $k \in \mathbb{N}$:

$$d + f^{(k)} \geq 2, \quad \tfrac{|u^{(k)} + f^{(k)}|}{d + f^{(k)}} \leq 1, \qquad \text{on} \quad \Omega_k. \tag{4.5.28}$$

II. Further estimates.

From now on we assume that $n \geq 5$. We denote by $\mathcal{S}(\Omega, C)$ the class of strictly convex C^∞-functions u, defined on Ω, such that

$$\inf_\Omega u(\xi) = 0, \qquad u|_{\partial\Omega} = C,$$

where C is a positive constant. Assume that $\Omega \subset \mathbb{R}^n$ is a normalized domain, and $u \in \mathcal{S}(\Omega, C)$ with $u(p) = 0$. We introduce the abbreviations:

$$\mathcal{A} := \max_\Omega \left\{ \exp\left\{ -\tfrac{m}{C-u} \right\} \frac{\rho^\alpha \Phi}{(d+f)^{\frac{2n\alpha}{n+2}}} \right\},$$

$$\mathcal{B} := \max_\Omega \left\{ \exp\left\{ -\tfrac{m}{C-u} + \tau \right\} \frac{(\gamma+2\alpha)\rho^\alpha}{(d+f)^{\frac{2n\alpha}{n+2}}} \right\},$$

where

$$\alpha := n+2, \qquad \text{for} \qquad n = 5 \text{ and } 6;$$

$$\alpha := \tfrac{(n+2)(n-3)}{2} + \tfrac{n-1}{4}, \qquad \text{for} \qquad n \geq 7,$$

and

$$m := 32(n+2)C, \qquad \tau := \epsilon \tfrac{\sum (x^k)^2}{(d+f)^2}.$$

From Lemma 4.5.4 we choose the constants ϵ and d such that

$$\tau < \tfrac{1}{30}, \qquad d + f \geq 2, \qquad \tfrac{|u+f|}{d+f} \leq 1$$

on the normalized domain Ω.
The following lemmas are important tools for the proof of the extension of Pogorelov's Theorem.

Lemma 4.5.5. *Let $\Omega \subset \mathbb{R}^n$ be a normalized domain, and $u \in \mathcal{S}(\Omega, C)$ with $u(p) = 0$ which satisfies the equation (4.5.1). Then there is a constant $d_1 > 0$, depending only on n and C, such that*

$$\mathcal{A} \leq d_1, \quad \mathcal{B} \leq d_1.$$

Proof. Again we consider the graph hypersurface, defined by the Legendre transform function f of u, together with the normalization by a constant transversal field such that we can apply the tools from sections 3.3.4 and 4.5.2 as before.

First step. We will prove the inequality $\mathcal{A} \leq 30\mathcal{B}$.
To this end, for $0 < \alpha \in \mathbb{R}$, recall the definitions of the functions ρ and Φ from section 1.4 and consider the following function

$$F := \exp\left\{ -\tfrac{m}{C-u} \right\} \frac{\rho^\alpha \Phi}{(d+f)^{\frac{2n\alpha}{n+2}}}$$

defined on Ω. Clearly, F attains its supremum at some interior point p^* of Ω. Thus, at p^*,

$$\frac{\Phi_{,i}}{\Phi} + \alpha\frac{\rho_{,i}}{\rho} - \frac{2n\alpha}{n+2}\cdot\frac{f_{,i}}{d+f} - \gamma u_{,i} = 0, \tag{4.5.29}$$

$$\frac{\Delta\Phi}{\Phi} - \frac{\sum(\Phi_{,i})^2}{\Phi^2} + \frac{n+2}{2}\alpha\Phi - \frac{2n\alpha}{n+2}\cdot\frac{\Delta f}{d+f}$$
$$+\frac{2n\alpha}{n+2}\cdot\frac{\sum(f_{,i})^2}{(d+f)^2} - \gamma'\sum(u_{,i})^2 - \gamma\Delta u \le 0, \tag{4.5.30}$$

where "," again denotes covariant derivation with respect to the Calabi metric $\mathfrak{H}$. In the calculation of (4.5.30) we used (4.5.8). Next we insert the relations (4.5.5), (4.5.6) and Proposition 4.5.3 into (4.5.30) and get:

$$\left[\frac{n+2}{2}\alpha + \frac{(n+2)^2}{n-1}\right]\Phi + \frac{1}{n-1}\cdot\frac{\sum(\Phi_{,i})^2}{\Phi^2} + \frac{n+2}{2}\cdot\gamma\cdot\frac{\sum u_{,i}\rho_{,i}}{\rho}$$
$$-n\alpha\frac{\sum f_{,i}\rho_{,i}}{(d+f)\rho} + \frac{2n\alpha}{n+2}\cdot\frac{\sum(f_{,i})^2}{(d+f)^2} + \frac{n^2-3n-10}{2(n-1)}\sum\frac{\Phi_{,i}}{\Phi}\cdot\frac{\rho_{,i}}{\rho}$$
$$-\gamma'\sum(u_{,i})^2 - n\gamma - \frac{2n\alpha}{n+2}\cdot\frac{n}{d+f} \le 0.$$

This and (4.5.29) give

$$\frac{1}{n-1}\sum\left[\gamma u_{,i} + \frac{2n\alpha}{n+2}\cdot\frac{f_{,i}}{d+f} - \alpha\frac{\rho_{,i}}{\rho}\right]^2 + \left[\frac{2(n+2)}{n-1}\alpha + \frac{(n+2)^2}{n-1}\right]\Phi$$
$$+\frac{(n+2)(n-3)}{n-1}\cdot\gamma\cdot\frac{\sum u_{,i}\rho_{,i}}{\rho} - \frac{4n\alpha}{n-1}\cdot\frac{\sum f_{,i}\rho_{,i}}{(d+f)\rho}$$
$$+\frac{2n\alpha}{n+2}\cdot\frac{\sum(f_{,i})^2}{(d+f)^2} - \gamma'\sum(u_{,i})^2 - n\gamma - \frac{2n^2\alpha}{n+2} \le 0. \tag{4.5.31}$$

Obey the inequality

$$\frac{|\sum u_{,i}f_{,i}|}{d+f} = \frac{|\sum\frac{\partial u}{\partial\xi_i}\cdot\frac{\partial f}{\partial x^k}u_{kj}u^{ij}|}{d+f} = \frac{|\sum\xi_i\cdot\frac{\partial u}{\partial\xi_i}|}{d+f} = \frac{|u+f|}{d+f} \le 1 \tag{4.5.32}$$

and insert it into (4.5.31):

$$\frac{1}{n-1}\cdot(\gamma)^2\sum(u_{,i})^2 + \left[\frac{4n^2\alpha^2}{(n+2)^2(n-1)} + \frac{2n\alpha}{n+2}\right]\cdot\frac{\sum(f_{,i})^2}{(d+f)^2}$$
$$+\frac{(n+2)(n-3)-2\alpha}{n-1}\gamma\frac{\sum u_{,i}\rho_{,i}}{\rho} + \frac{(\alpha+n+2)^2}{n-1}\Phi - \frac{2n^2\alpha}{n+2}$$
$$-\frac{4n\alpha(\alpha+n+2)}{(n-1)(n+2)}\cdot\frac{\sum f_{,i}\rho_{,i}}{(d+f)\rho} - \gamma'\sum(u_{,i})^2 - \left(n + \frac{4n\alpha}{(n-1)(n+2)}\right)\gamma \le 0. \tag{4.5.33}$$

We discuss two cases:
(i) For the dimension $n = 5$ and 6 we choose $\alpha = n+2$. In this case it is easy to check that

$$\frac{4n^2\alpha^2}{(n+2)^2(n-1)} + \frac{2n\alpha}{n+2} = \frac{2n(3n-1)}{n-1}.$$

Using the inequality of Schwarz we get

$$\frac{(n+2)(n-3)-2\alpha}{n-1}\gamma\frac{\sum u_{,i}\rho_{,i}}{\rho} \le \frac{1}{2(n-1)}\gamma^2\sum(u_{,i})^2 + \frac{(n+2)^2(n-5)^2}{2(n-1)}\Phi,$$

$$\frac{4n\alpha(\alpha+n+2)}{(n-1)(n+2)}\frac{\sum f_{,i}\rho_{,i}}{(d+f)\rho} \le \frac{2n(3n-1)}{n-1}\frac{\sum(f_{,i})^2}{(d+f)^2} + \frac{8n(n+2)^2}{(3n-1)(n-1)}\Phi.$$

(ii) For $n \geq 7$ we choose $\alpha = \frac{(n+2)(n-3)}{2} + \frac{n-1}{4}$. Then it is easy to check that

$$\frac{4n^2\alpha^2}{(n+2)^2(n-1)} + \frac{2n\alpha}{n+2} = \frac{4n^2\alpha^2}{(n+2)^2(n-1)}\left(1 + \frac{(n+2)(n-1)}{2n\alpha}\right) > \frac{4n^2\alpha^2}{(n+2)(n^2-1)}.$$

Again we use the inequality of Schwarz and have

$$\frac{1}{2}\gamma\frac{\sum u_{,i}\rho_{,i}}{\rho} \leq \frac{1}{2(n-1)}\gamma^2\sum(u_{,i})^2 + \frac{n-1}{8}\Phi,$$

$$\frac{n(2n+5)\alpha}{n+2}\frac{\sum f_{,i}\rho_{,i}}{(d+f)\rho} \leq \frac{4n^2\alpha^2}{(n+2)(n^2-1)}\frac{\sum(f_{,i})^2}{(d+f)^2} + \frac{(2n+5)^2(n^2-1)}{16(n+2)}\Phi.$$

After the separate discussions of $n = 5, 6$ in (i) and $n \geq 7$ in (ii), we return to (4.5.33) for all $n \geq 5$. Note that $\frac{1}{2(n-1)}\gamma^2 \geq \gamma'$. From (4.5.33) and the inequalities above we get:

$$\frac{1}{10}\frac{(n+2)^2}{n-1}\Phi - \left(n + \frac{4n\alpha}{(n-1)(n+2)}\right)\gamma - \frac{2n^2\alpha}{n+2} \leq 0.$$

This finally gives the claim in the first step.

Second step. Now we consider the following function

$$\tilde{F} := \exp\left\{-\frac{m}{C-u} + \tau\right\} \cdot \frac{(\gamma+2\alpha)\rho^\alpha}{(d+f)^{\frac{2n\alpha}{n+2}}}$$

defined on Ω. Clearly, $\tilde{F}$ attains its supremum at some interior point $q^* \in \Omega$. Thus, at q^*,

$$-\gamma u_{,i} + \frac{\gamma' u_{,i}}{\gamma+2\alpha} + \tau_{,i} + \alpha\cdot\frac{\rho_{,i}}{\rho} - \frac{2n\alpha}{n+2}\cdot\frac{f_{,i}}{d+f} = 0, \tag{4.5.34}$$

and with the definition $\gamma'' := \frac{6m}{(C-u)^4}$

$$\left(\frac{\gamma''}{\gamma+2\alpha} - \frac{\gamma'^2}{(\gamma+2\alpha)^2} - \gamma'\right)\sum(u_{,i})^2 + \left(\frac{\gamma'}{\gamma+2\alpha} - \gamma\right)\Delta u$$
$$+ \Delta\tau + \frac{n+2}{2}\alpha\Phi - \frac{2n\alpha}{n+2}\left(\frac{\Delta f}{d+f} - \frac{\sum(f_{,i})^2}{(d+f)^2}\right) \leq 0. \tag{4.5.35}$$

We use the inequality of Schwarz and the calculation of the Laplacian in section 3.3.4 to derive the following two relations:

$$\sum(\tau_{,i})^2 \leq 8\epsilon\tau\frac{\sum f^{ii}}{(d+f)^2} + 8\tau^2\frac{\sum(f_{,i})^2}{(d+f)^2}, \tag{4.5.36}$$

$$\begin{aligned}
\Delta\tau &= \epsilon\,\frac{\Delta(\sum(x^k)^2)}{(d+f)^2} - 4\epsilon\,\frac{\mathfrak{H}(\mathrm{grad}(\sum(x^k)^2), \mathrm{grad} f)}{(d+f)^3} \\
&\quad - 2\epsilon\cdot\frac{\sum(x^k)^2\,\Delta f}{(d+f)^3} + 6\epsilon\,\frac{\sum(x^k)^2\sum(f_{,i})^2}{(d+f)^4} \\
&= \frac{\epsilon}{(d+f)^2}\left[2\sum f^{ii} + \frac{n+2}{2}\mathfrak{H}\left(\mathrm{grad}\ln\rho, \mathrm{grad}\left(\sum(x^k)^2\right)\right)\right] \\
&\quad - \frac{4\epsilon}{(d+f)^2}\frac{\mathfrak{H}\left(\mathrm{grad}(\sum(x^k)^2), \mathrm{grad} f\right)}{d+f} + 6\epsilon\,\frac{\sum(x^k)^2\cdot\sum(f_{,i})^2}{(d+f)^4} \\
&\quad - 2n\epsilon\,\frac{\sum(x^k)^2}{(d+f)^3} - (n+2)\epsilon\,\frac{\sum(x^k)^2\,\mathfrak{H}(\mathrm{grad}\ln\rho, \mathrm{grad} f)}{(d+f)^3} \\
&\geq \frac{\epsilon}{(d+f)^2}\sum f^{ii} - 27\tau\,\frac{\sum(f_{,i})^2}{(d+f)^2} - \frac{3(n+2)^2}{4}\tau\Phi - 2n\tau.
\end{aligned} \tag{4.5.37}$$

Note that $\frac{(n+2)^2}{2} > \alpha \geq n+2$ and (4.5.34); we obtain

$$\begin{aligned}
\Phi =& \tfrac{1}{\alpha^2} \sum \left(-\gamma u_{,i} + \tfrac{\gamma' u_{,i}}{\gamma+2\alpha} + \tau_{,i} - \tfrac{2n\alpha}{n+2} \tfrac{f_{,i}}{d+f} \right)^2 \\
\geq& \tfrac{1}{2\alpha^2} \sum \left(-\gamma u_{,i} + \tfrac{\gamma' u_{,i}}{\gamma+2\alpha} - \tfrac{2n\alpha}{n+2} \tfrac{f_{,i}}{d+f} \right)^2 - \tfrac{1}{\alpha^2} \sum (\tau_{,i})^2 \\
\geq& \tfrac{\gamma^2}{4\alpha^2} \sum (u_{,i})^2 + \tfrac{n^2}{(n+2)^2} \tfrac{\sum (f_{,i})^2}{(d+f)^2} - \tfrac{1}{2\alpha^2} \tfrac{\gamma'^2 \sum (u_{,i})^2}{(\gamma+2\alpha)^2} - \tfrac{1}{\alpha^2} \sum (\tau_{,i})^2 - \gamma,
\end{aligned}$$

where we use the fact (4.5.32).
Now we insert (4.5.5), (4.5.6), (4.5.34) and (4.5.36-37) into (4.5.35) and use the inequality of Schwarz; this gives

$$\tfrac{\epsilon}{2} \cdot \tfrac{\sum f^{ii}}{(d+f)^2} - a_0 \Phi - a_1 \gamma - 3n\alpha \leq 0$$

for some constants $a_0 > 0$ and $a_1 > 0$, depending only on n. Since $\sum f^{ii} \geq n\rho^{\frac{n+2}{n}}$, we get

$$\frac{\rho^{\frac{n+2}{n}}}{(d+f)^2} \leq \tfrac{1}{\epsilon} \left(a_0 \Phi + 2a_1 \gamma + 6\alpha \right).$$

It follows that

$$\mathcal{B}^{1+\frac{n+2}{n\alpha}} \leq a_2 \mathcal{A} + a_3 \mathcal{B}$$

for some positive constants a_2 and a_3. In the first step we proved $\mathcal{A} \leq 30 \cdot \mathcal{B}$, this finally gives

$$\mathcal{B} \leq d_1, \qquad \mathcal{A} \leq d_1$$

for some constant d_1 depending only on C and n. This proves Lemma 4.5.5. ∎

For the next Lemma we introduce the following notation and assumptions. Let $u(\xi) \in \mathcal{S}(\Omega, C)$ (see section 4.5. II). Consider the function

$$F^{\sharp} := \exp\left\{ -\tfrac{m}{C-u} + \tau^{\sharp} \right\} \cdot Q \cdot \|\mathrm{grad}\, K\|^2, \tag{4.5.38}$$

where $Q > 0$, $\tau^{\sharp} > 0$ and K are smooth functions defined on the closure $\overline{\Omega}$. $F^{\sharp}$ attains its supremum at an interior point p^*. We choose a local orthonormal frame field on M such that, at p^*, the function K satisfies $K_{,1} = \|\mathrm{grad}\, K\|$, and $K_{,i} = 0$, for all $i > 1$.

Lemma 4.5.6. *Under the forgoing notations and assumptions, at the point p^*, we have the following estimates*

$$\begin{aligned}
&2 \left(\tfrac{1}{n-1} - \delta - 1 \right) (K_{,11})^2 + 2 \sum K_{,j} (\Delta K)_{,j} \\
&+ 2(1-\delta) \sum A^2_{ml1} (K_{,1})^2 - \tfrac{(n+2)^2}{8\delta} \Phi (K_{,1})^2 - \tfrac{2}{\delta(n-1)^2} (\Delta K)^2 \\
&+ \left[-\gamma' \sum (u_{,i})^2 - \gamma \Delta u + \Delta \tau^{\sharp} + \tfrac{\Delta Q}{Q} - \tfrac{\sum (Q_{,i})^2}{Q^2} \right] (K_{,1})^2 \leq 0
\end{aligned} \tag{4.5.39}$$

for any positive number $0 < \delta < 1$.

Proof. We can assume that $\|\mathrm{grad}\, K\|(p^*) > 0$. Then, at p^*,

$$F^\sharp_{,i} = 0 \qquad \text{and} \qquad \sum F^\sharp_{,ii} \leq 0.$$

Using the definition of $F^\sharp$, we calculate both expressions explicitly and get

$$\left(-\gamma u_{,i} + \tau^\sharp_{,i} + \tfrac{Q_{,i}}{Q}\right) \sum (K_{,j})^2 + 2\sum K_{,j} K_{,ji} = 0, \tag{4.5.40}$$

$$\begin{aligned} 2\sum (K_{,ij})^2 + 2\sum K_{,j} K_{,jii} + 2\sum \left(-\gamma u_{,i} + \tau^\sharp_{,i} + \tfrac{Q_{,i}}{Q}\right) K_{,j} K_{,ji} \\ + \left[-\gamma' \sum (u_{,i})^2 - \gamma \Delta u + \Delta \tau^\sharp + \tfrac{\Delta Q}{Q} - \tfrac{\sum (Q_{,i})^2}{Q^2}\right] (K_{,1})^2 \leq 0. \end{aligned} \tag{4.5.41}$$

Let us simplify (4.5.41); (4.5.40) implies

$$2K_{,1i} = \left(\gamma u_{,i} - \tau^\sharp_{,i} - \tfrac{Q_{,i}}{Q}\right) K_{,1}. \tag{4.5.42}$$

Apply the inequality of Schwarz to (4.5.41):

$$\begin{aligned} 2\sum (K_{,ij})^2 &\geq 2(K_{,11})^2 + \tfrac{2}{n-1}(\Delta K - K_{,11})^2 + 4\sum_{i>1} (K_{,1i})^2 \\ &\geq 2\left(\tfrac{n}{n-1} - \delta\right)(K_{,11})^2 + 4\sum_{j>1} (K_{,1j})^2 - \tfrac{2}{\delta(n-1)^2}(\Delta K)^2 \end{aligned} \tag{4.5.43}$$

for any $\delta > 0$. Inserting (4.5.42) and (4.5.43) into (4.5.41), this gives:

$$\begin{aligned} 2\left(\tfrac{1}{n-1} - \delta - 1\right)(K_{,11})^2 + 2\sum K_{,j} K_{,jii} - \tfrac{2}{\delta(n-1)^2} \cdot (\Delta K)^2 \\ + \left[-\gamma' \sum (u_{,i})^2 - \gamma \Delta u + \Delta \tau^\sharp + \tfrac{\Delta Q}{Q} - \tfrac{\sum (Q_{,i})^2}{Q^2}\right] \cdot (K_{,1})^2 \leq 0. \end{aligned} \tag{4.5.44}$$

For the third order terms $K_{,jii}$ we apply the Ricci identity with respect to the Calabi metric and obtain:

$$\begin{aligned} 2\sum K_{,j} K_{,jii} &= 2\sum K_{,j} (\Delta K)_{,j} + 2R_{11}(K_{,1})^2 \\ &= 2\sum K_{,j}(\Delta K)_{,j} + 2\sum A^2_{ml1}(K_{,1})^2 - (n+2)\sum A_{11k}\tfrac{\rho_k}{\rho} \cdot (K_{,1})^2 \\ &\geq 2\sum K_{,j}(\Delta K)_{,j} + 2(1-\delta)\sum A^2_{ml1}(K_{,1})^2 - \tfrac{(n+2)^2}{8\delta}\Phi (K_{,1})^2. \end{aligned}$$

We insert the last inequality into (4.5.44); this proves the assertion of Lemma 4.5.6. ∎

Lemma 4.5.7. *Let $\Omega \subset \mathbb{R}^n$ be a normalized domain, and $u \in \mathcal{S}(\Omega, C)$ with $u(p) = 0$ which satisfies the equation (4.5.1). Then there is a constant $d_2 > 0$, depending only on n and C, such that*

$$\exp\left\{-\tfrac{64(n-1)C}{C-u}\right\} \cdot \frac{\rho^\alpha \sum u_{ii}}{(d+f)^{\frac{2n\alpha}{n+2}+2}} \leq d_2$$

on Ω, where α is the constant in Lemma 4.5.5.

Proof. Recall the definition of the function $F^\sharp$ in (4.5.38) and choose now the following explicit functions

$$\tau^\sharp := \epsilon \cdot \tfrac{\sum (x^k)^2}{(d+f)^2}, \qquad K := x^1, \qquad Q := \frac{\rho^\alpha}{(d+f)^{\frac{2n\alpha}{n+2}+2}}.$$

First step. We want to give a lower estimate for the following expression:

$$2\sum K_{,j}(\Delta K)_{,j} + 2(1-\delta)\sum A_{mli}A_{mlj}K_{,i}K_{,j}.$$

For this purpose we recall the expression for the Laplacian from the end of section 3.3.4 and the fact that, in local terms, the Hessians of the graph functions f and u, namely f_{ij} and u_{ij}, give the Calabi metric, resp., and f^{ij} and u^{ij} give the inverse matrices, resp. Now we calculate the following two relations

$$\Delta K = \tfrac{n+2}{2}\cdot \mathfrak{H}(\text{grad}\ln\rho\,,\,\text{grad}\,K),$$

$$\begin{aligned}2\sum K_{,j}(\Delta K)_{,j} &= (n+2)\left(\tfrac{\rho_{,11}}{\rho}\cdot(K_{,1})^2 - \tfrac{(\rho_{,1})^2}{\rho^2}\cdot(K_{,1})^2 + \sum K_{,1i}\,K_{,1}\cdot\tfrac{\rho_{,i}}{\rho}\right)\\ &\geq (n+2)\sum\tfrac{\rho_{,ij}}{\rho}\cdot K_{,i}\,K_{,j} - \delta\sum(K_{,1i})^2 - \tfrac{(n+2)^2+1}{4\delta}\,\Phi\,(K_{,1})^2\end{aligned}$$

for $\delta \leq \frac{1}{4(n+2)}$. In terms of the coordinates $\xi_1,...,\xi_n$ we calculate the sums

$$\sum(K_{,ij})^2 \qquad\text{and}\qquad \sum A^2_{ml1}(K_{,1})^2.$$

We recall the expression for the Christoffel symbols of the Levi-Civita connection: $\Gamma^k_{ij} = \frac{1}{2}\sum u^{kl}u_{lij}$, then

$$K_{,ij} = u_{1ij} - \tfrac{1}{2}\sum u_{1k}\,u^{kl}\,u_{lij} = \tfrac{1}{2}u_{1ij},$$

$$\sum(K_{,ij})^2 = \tfrac{1}{4}\sum u^{ik}\,u^{jl}\,u_{1ij}\,u_{1kl},$$

$$\sum A^2_{ml1}(K_{,1})^2 = \tfrac{1}{4}\sum u^{ik}\,u^{jl}\,u_{ijp}\,u_{klq}\,u^{pr}\,u_{1r}\,u^{qs}\,u_{1s} = \sum(K_{,ij})^2.$$

We apply (4.5.7) and calculate in terms of the coordinates $x^1,...,x^n$:

$$\tfrac{\rho_{ij}}{\rho} = \tfrac{\rho_i}{\rho}\cdot\tfrac{\rho_j}{\rho} \quad\text{and}\quad \tfrac{\rho_{,ij}}{\rho} = \tfrac{\rho_i}{\rho}\cdot\tfrac{\rho_j}{\rho} + \sum A^k_{ij}\,\tfrac{\rho_k}{\rho}.$$

This gives

$$(n+2)\sum\tfrac{\rho_{,ij}}{\rho}\,K_{,i}\,K_{,j} \;\leq\; \delta\sum(K_{,ij})^2 + \tfrac{(n+2)^2+1}{4\delta}\cdot\Phi(K_{,1})^2 \tag{4.5.45}$$

and

$$\begin{aligned}2\sum K_{,j}(\Delta K)_{,j} \;+\; 2(1-\delta)\sum A_{mli}\,A_{mlj}\,K_{,i}\,K_{,j}&\\ \geq (2-4\delta)\sum(K_{,ij})^2 - \tfrac{(n+2)^2+1}{2\delta}\cdot\Phi\,(K_{,1})^2.&\end{aligned} \tag{4.5.46}$$

Second step. In (4.5.41) the following expression appears (on the left hand side), and we calculate:

$$\tfrac{\Delta Q}{Q} - \tfrac{\sum(Q_{,i})^2}{Q^2} \geq -\tfrac{(n\alpha+n+2)(n+2)}{8}\cdot\Phi - 2n(\alpha+1). \tag{4.5.47}$$

Next we calculate (4.5.42) and use (4.5.32):

$$\begin{aligned}\sum(K_{,1i})^2 &= \tfrac{1}{4}\sum\left[\gamma\,u_{,i} - \alpha\cdot\tfrac{\rho_{,i}}{\rho} + \left(\tfrac{2n\alpha}{n+2}+2\right)\cdot\tfrac{f_{,i}}{d+f} - \tau^\sharp_{,i}\right]^2\cdot(K_{,1})^2\\ &\geq \left(\tfrac{1}{16}\sum\left[\gamma\,u_{,i} + \left(\tfrac{2n\alpha}{n+2}+2\right)\tfrac{f_{,i}}{d+f}\right]^2 - \tfrac{1}{8}\alpha^2\,\Phi - \tfrac{1}{4}\sum(\tau^\sharp_{,i})^2\right)(K_{,1})^2\\ &\geq \tfrac{1}{16}\left[\gamma^2\sum(u_{,i})^2 + \tfrac{4n^2\alpha^2}{(n+2)^2}\tfrac{\sum(f_{,i})^2}{(d+f)^2}\right](K_{,1})^2\\ &\qquad - \tfrac{1}{8}\alpha^2\Phi(K_{,1})^2 - \tfrac{1}{4}(K_{,1})^2\sum(\tau^\sharp_{,i})^2 - a_4\gamma(K_{,1})^2\end{aligned} \tag{4.5.48}$$

for some positive constant a_4.
In the definition of the function $F^\sharp$ there appear constants m and C, while the constant $\frac{1}{4(n+2)} > \delta > 0$ in Lemma 4.5.6 is arbitrary so far and appears in (4.5.45); now we choose

$$\delta = \frac{1}{6(n+2)} \quad \text{and} \quad m = 64(n-1)C,$$

insert (4.5.36-37) and (4.5.46-48) into (4.5.39) and apply again the inequality of Schwarz; we get:

$$\frac{\epsilon}{2} \cdot \frac{\sum f^{ii}}{(d+f)^2} - a_5\Phi - a_6\gamma - a_7 \leq 0. \tag{4.5.49}$$

The constants $a_4, ..., a_7$, that appear in the foregoing calculations, depend only on the dimension n. Note that

$$\sum f^{ii} \geq u_{11} = (K_{,1})^2.$$

Using Lemma 4.5.5 we obtain that

$$\exp\left\{-\frac{m}{C-u}\right\} \cdot \frac{\rho^\alpha\, u_{11}}{(d+f)^{\frac{2n\alpha}{n+2}+2}} \leq d_2$$

for some constant d_2 that depends only on n and C. Similar inequalities for u_{ii} are true. Thus the proof of Lemma 4.5.7 is complete. ■

To sum up, from Lemmas 4.5.5 and 4.5.7, we get the following estimates using our notational agreement from the beginning of section 4.5.6.

Proposition 4.5.8. *Let $\Omega \subset \mathbb{R}^n$ be a smooth normalized convex domain and 0 the center of Ω. Let u be a strictly convex C^∞ function defined on Ω. Assume that*

$$\inf_\Omega u = 0, \qquad u = C > 0 \quad \text{on } \partial\Omega$$

and u satisfies the PDE (4.5.1). Then there exists a constant

$$\alpha := n+2, \qquad \text{for} \qquad n = 5 \text{ and } 6;$$

$$\alpha := \frac{(n+2)(n-3)}{2} + \frac{n-1}{4}, \qquad \text{for} \qquad n \geq 7,$$

such that

$$\frac{\rho}{(d+f)^{\frac{2n}{n+2}}} \leq d_3, \quad \frac{\rho^\alpha \Phi}{(d+f)^{\frac{2n\alpha}{n+2}}} \leq d_3, \quad \frac{\rho^\alpha \sum u_{ii}}{(d+f)^{\frac{2n\alpha}{n+2}+2}} \leq d_3$$

on $S_u(\frac{C}{2}) := \{\xi \in \mathbb{R}^n \mid u \leq \frac{C}{2}\}$ for some constants $d > C$ and $d_3 > 0$, where d_3 depends only on n and C.

4.5.7 *Proof of Theorem 4.5.1 - $n \geq 5$*

Now we come back to the Monge-Ampère equation in the version (4.5.3) for the Legendre transform function f of the strictly convex function $u : \mathbb{R}^n \to \mathbb{R}$. We recall our notational agreement from section 4.5.6. We equip $\mathbb{R}^n$ with the canonical Euclidean structure.

Step 1. Let $p \in \mathbb{R}^n$ be any point. By subtraction of a linear function we may assume that u satisfies

$$\operatorname{grad} u(p) = 0, \qquad u(\xi) \geq u(p) = 0, \quad \forall \xi \in \mathbb{R}^n.$$

Choose a sequence $\{C_k\}$ of positive numbers such that $C_k \to \infty$ as $k \to \infty$. In analogy to section 4.5.4 we have, for any C_k, that the level set

$$S_u(p, C_k) := \{\xi \mid u(\xi) < C_k\}$$

is a bounded convex domain. Define the sequence $u^{(k)} : \mathbb{R}^n \to \mathbb{R}$ by

$$u^{(k)}(\xi) := \tfrac{u(\xi)}{C_k}, \qquad k = 1, 2, \ldots$$

By q_k denote the center of mass of $S_u(p, C_k)$. We repeat an argument from the beginning of section 4.5.6, see [37]: For each k there exists the unique minimum ellipsoid E_k of $S_u(p, C_k)$, centered at q_k, such that

$$n^{-\frac{3}{2}} E_k \subset S_u(p, C_k) \subset E_k.$$

For fixed k define a linear transformation $T_k : \mathbb{R}^n \to \mathbb{R}^n$ by

$$T_k : \tilde{\xi}_i = \sum a_i^j \xi_j + b_i$$

such that

$$T_k(q_k) = 0, \quad T_k(E_k) = B_1(0).$$

Then

$$B_{n^{-\frac{3}{2}}}(0) \subset \Omega_k := T_k(S_u(p, C_k)) \subset B_1(0).$$

Thus we obtain a sequence of convex functions

$$\tilde{u}^{(k)}(\tilde{\xi}) := u^{(k)}\left(\sum b_1^j(\tilde{\xi}_j - b_j), \ldots, \sum b_n^j(\tilde{\xi}_j - b_j)\right),$$

where $\tilde{u}^{(k)} : \Omega_k \to \mathbb{R}$, in terms of the new coordinates $\tilde{\xi}_j$, and where $(b_i^j) = (a_i^j)^{-1}$.

Step 2. We simplify the notation and proceed with $u^{(k)} : \Omega_k \to \mathbb{R}$ instead of $\tilde{u}^{(k)}$, and ξ instead of $\tilde{\xi}$. Considering appropriate subsequences, we then may assume that the domains Ω_k converge to a convex domain Ω and the functions $u^{(k)}(\xi)$ converge to a convex function, locally uniformly in Ω. For any fixed function $u^{(k)}$ consider its Legendre transformation with coordinates

$$x^i = \tfrac{\partial u^{(k)}}{\partial \xi_i},$$

and its Legendre transform functions

$$f^{(k)}(x^1, \ldots, x^n) = \sum \xi_i \tfrac{\partial u^{(k)}}{\partial \xi_i} - u^{(k)}(\xi_1, \ldots, \xi_n), \qquad (\xi_1, \ldots, \xi_n) \in \Omega_k,$$

defined on the associated Legendre transform domains

$$\Omega_k^* := \left\{(x^1, ..., x^n) \mid x^i = \frac{\partial u^{(k)}}{\partial \xi_i}\right\}.$$

Obviously, any $f^{(k)}$ satisfies the PDE of type (4.5.3), therefore there exist real constants $d_1^{(k)}, ..., d_n^{(k)}, d_0^{(k)}$ such that

$$\det\left(\frac{\partial^2 f^{(k)}}{\partial x^i \partial x^j}\right) = \exp\left\{\sum d_i^{(k)} x^i + d_0^{(k)}\right\}. \tag{4.5.50}$$

Step 3. Now we will use our technical tools from the foregoing section, namely for each function $u^{(k)}$ we apply Proposition 4.5.8, where we set $C = 1$. This way we get the following uniform estimates

$$\frac{\rho^{(k)}}{(d+f^{(k)})^{\frac{2n}{n+2}}} \leq d_4, \qquad \frac{(\rho^{(k)})^\alpha \cdot \Phi^{(k)}}{(d+f^{(k)})^{\frac{2n\alpha}{n+2}}} \leq d_4, \qquad \frac{(\rho^{(k)})^\alpha \sum u_{ii}^{(k)}}{(d+f^{(k)})^{\frac{2n\alpha}{n+2}+2}} \leq d_4$$

on $S_{u^{(k)}}(T^k(p), \frac{1}{2})$ for the same appropriate constant $d_4 > 0$, where α is a positive constant.

Step 4. Let $B_R(0)$ be a Euclidean ball of radius R such that $S_{u^{(k)}}(T^k(p), \frac{1}{2}) \subset B_{\frac{1}{2}R}(0)$, for all k. The comparison theorem for the normal mapping (see section 1.3) yields

$$\bar{B}_\delta^*(0) \subset \Omega_k^*$$

for every k, where $\delta = \frac{1}{2R}$ and $\bar{B}_\delta^*(0) = \{x \mid \sum (x^i)^2 \leq \delta^2\}$. Note that $u^k(T^k(p)) = 0$ and its image under the normal mapping is $(x^1, ..., x^n) = 0$. Restricting to $\bar{B}_\delta^*(0)$, we have

$$-R' \leq f^{(k)} = \sum \xi_i x^i - u^{(k)} \leq R',$$

where $R' = \frac{1}{R} + 1$. Thus the sequence $\{f^{(k)}\}$ locally uniformly converges to a convex function f^∞ on $\bar{B}_\delta^*(0)$; moreover, applying the foregoing to the terms of the form $(d + f^{(k)})$, we conclude that there exist uniform estimates

$$\rho^{(k)} \leq d_5, \qquad (\rho^{(k)})^\alpha \Phi^{(k)} \leq d_5, \qquad (\rho^{(k)})^\alpha \sum u_{ii}^{(k)} \leq d_5 \tag{4.5.51}$$

on $\bar{B}_\delta^*(0)$ for the same appropriate constant $d_5 > 0$.

Step 5. Below we apply the following Lemma to any function $f^{(k)}$ from step 4.
Lemma 4.5.9. *Let $f : x \mapsto f(x)$ be a smooth strictly convex function defined on $\bar{B}_\delta^*(0)$ satisfying*

$$-R' \leq f \leq R'.$$

Then there exists a point $p^* \in \bar{B}_\delta^*(0)$ such that, at p^*,

$$\frac{1}{\rho} < \left(\frac{4R'}{\delta^2}\right)^{\frac{n}{n+2}} 2^{\frac{n+1}{n+2}} =: d_6,$$

where ρ was defined in terms of f according to our notational agreement above.

Proof. Assume the assertion in Lemma is not correct. Then

$$\tfrac{1}{\rho} \geq d_6 \quad \text{on} \quad \bar{B}^*_\delta(0).$$

It follows that

$$\det(f_{ij}) \geq (d_6)^{n+2} \quad \text{on} \quad \bar{B}^*_\delta(0).$$

Define the function

$$F^{\sharp\sharp}(x) := \left(\tfrac{(d_6)^{n+2}}{2^{n+1}}\right)^{\frac{1}{n}} \left(\sum (x^i)^2 - \delta^2\right) + 2R' \quad \text{on} \quad \bar{B}^*_\delta(0).$$

Then

$$\det(F^{\sharp\sharp}_{ij}) = \tfrac{1}{2} \cdot (d_6)^{n+2} < \det(f_{ij}) \quad \text{on} \quad \bar{B}^*_\delta(0)$$

and

$$F^{\sharp\sharp}(x) \geq f(x) \quad \text{on} \quad \partial\bar{B}^*_\delta(0).$$

The comparison principle for Monge-Ampère equations (see section 1.3) implies

$$F^{\sharp\sharp}(x) \geq f(x) \quad \text{on} \quad \bar{B}^*_\delta(0).$$

On the other hand we have

$$F^{\sharp\sharp}(0) = -\left(\tfrac{d_6^{n+2}}{2^{n+1}}\right)^{\frac{1}{n}} \delta^2 + 2R' = -2R' < f(0).$$

This is a contradiction, thus Lemma 4.5.9 is proved. ∎

Step 6. We are now ready for the final steps of the proof of the extension of Pogorelov's Theorem. From Lemma 4.5.9 and the uniform estimates in (4.5.51) it follows that, for fixed k, there exists a point $p_k \in \bar{B}^*_\delta(0)$ such that the four functions $\rho^{(k)}$, $\frac{1}{\rho^{(k)}}$, $\Phi^{(k)}$ and $\sum_i u^{(k)}_{ii}$ are uniformly bounded at p_k. Therefore there are constants

$$0 < \lambda_* \leq \lambda^* < \infty$$

that are independent of k, such that we get the following lower and upper bound for the eigenvalues $\nu^{(k)}$ of the matrix $(f^{(k)}_{ij})$ at p_k:

$$\lambda_* \leq \nu^{(k)}(p_k) \leq \lambda^*.$$

Since each $f^{(k)}$ satisfies the PDE (4.5.50), we have

$$\Phi^{(k)} = \sum f^{(k)ij} (\ln \rho^{(k)})_i (\ln \rho^{(k)})_j = \tfrac{1}{(n+2)^2} \sum f^{(k)ij} d^{(k)}_i d^{(k)}_j,$$

where the coefficients $d^{(k)}_i$ appeared in the PDE for $f^{(k)}$. From this it follows that

$$\sum (d^{(k)}_i)^2 \leq (n+2)^2 \lambda^* \Phi^{(k)}(p_k) \leq d_7$$

for some constant $d_7 > 0$. Thus

$$\|\text{grad } \ln \rho^{(k)}\|^2_E = \sum \left(\tfrac{\partial \ln \rho^{(k)}}{\partial x^i}\right)^2 = \tfrac{1}{(n+2)^2} \sum (d^{(k)}_i)^2 \leq d_7, \tag{4.5.52}$$

where $\|\cdot\|_E$ denotes the norm with respect to the Euclidean metric.

Step 7. For any unit speed geodesic, starting from p_k, it follows that

$$\left|\frac{d\ln\rho^{(k)}}{ds}\right| \leq \|\text{grad}\ln\rho^{(k)}\|_E \leq d_7. \tag{4.5.53}$$

Thus, for any point $q \in B^*_\delta(0)$, we have

$$\rho^{(k)}(p_k)\cdot\exp\{-|q-p_k|d_7\} \leq \rho^{(k)}(q) \leq \rho^{(k)}(p_k)\cdot\exp\{|q-p_k|d_7\}. \tag{4.5.54}$$

In particular, we choose q to be the point with coordinates $x^i = 0$ for all $i \geq 1$; that are the x-coordinates of the point p as considered in the beginning of Step 4. From the inequalities in (4.5.51) it follows that

$$\Phi^{(k)}(q) = \frac{\|\text{grad}\,\rho^{(k)}\|^2}{(\rho^{(k)})^2}(q) \leq d_8 \tag{4.5.55}$$

for some constant $d_8 > 0$ that is independent of k.

Step 8. Now assume that $\Phi(p) \neq 0$, then a direct calculation gives

$$\Phi^{(k)}(q) = C_k\Phi(p) \to \infty, \quad \text{as} \quad k \to \infty.$$

This contradicts (4.5.55), thus

$$\Phi(p) = 0.$$

Since p is arbitrary we conclude that $\Phi = 0$ everywhere. Consequently we arrive at the PDE

$$\det\left(\frac{\partial^2 u}{\partial\xi_i\partial\xi_j}\right) = const > 0.$$

Thus M is a Euclidean complete parabolic affine hypersphere and therefore an elliptic paraboloid. The proof of the extension of Pogorelov's Theorem is now complete. ■

4.6 A Cubic Form Differential Inequality with its Applications

The following well-known result of E. Calabi is related to the Theorem of Jörgens-Calabi-Pogorelov from section 4.4. There we assumed Euclidean completeness, while the following Theorem states that every affine complete, parabolic affine hypersphere is an elliptic paraboloid [19]. From an analytic point of view this result can be restated as follows:

Theorem 4.6.1. *Let f be a smooth, strictly convex solution of*

$$\det\left(\frac{\partial^2 f}{\partial x^i\partial x^j}\right) = 1 \quad on \quad \Omega$$

and $M = \{(x, f(x)) \mid x \in \Omega\}$ be the graph defined by f. If M is affine complete then f must be a quadratic polynomial.

Here we give the following generalization. For the definition of an α-*metric* we refer to section 3.3.5.

Theorem 4.6.2. [62]. *Let $f(x^1, ..., x^n)$ be a strictly convex C^∞-function defined on a domain $\Omega \subset \mathbb{R}^n$ satisfying the PDE (4.5.2). If $\alpha \neq n+2$ and the graph hypersurface $M = \{(x, f(x))\}$ is complete with respect to the α-metric then M must be an elliptic paraboloid.*

Remark. In case $\alpha = n+2$, the foregoing theorem is wrong. We give a counter example: the graph

$$h(x^1, ..., x^n) = \exp\{x^1\} + \sum_{i=2}^{n} (x^i)^2$$

is an $(n+2)$-complete solution of the PDE (4.5.2).

Remark. In case $\alpha = 0$, we get the following theorem [47].

Theorem. *Let M be an affine Kähler manifold. If the Hessian metric of M is affine Kähler-Ricci flat and complete, then M must be $\mathbb{R}^n/\Gamma$, where Γ is a subgroup of the group of isometries which acts freely and properly discontinuously on $\mathbb{R}^n$.*

4.6.1 Calculation of ΔJ in terms of the Calabi metric

In the following we will use the Calabi metric to calculate. Recall the definition of the functions ρ and Φ from section 3.3.4:

$$\varphi := -\tfrac{1}{2} \ln \det(f_{ij}) = \tfrac{n+2}{2} \ln \rho, \quad \Phi := \tfrac{\|\text{grad}\, \rho\|^2}{\rho^2} = \tfrac{4}{(n+2)^2} \parallel \text{grad}\, \varphi \parallel^2 .$$

Lemma 4.6.3. *Consider a locally strongly convex graph hypersurface with Calabi's normalization. In terms of the Calabi metric, the Laplacian of the relative Pick invariant J (see section 3.3.4) satisfies*

$$\Delta J \geq \tfrac{2}{n(n-1)} \sum A_{ijk}\, \varphi_{,ijk} + \tfrac{2}{n(n-1)} \parallel \nabla A \parallel^2 + 2J^2 - \tfrac{(n+2)^4}{4} \Phi^2;$$

here ∇A denotes the covariant derivative of A, and ∇ denotes the Levi-Civita connection $\nabla = \nabla(\mathfrak{H})$.

Proof. Choose a locally $\mathfrak{H}$-orthonormal frame field $x; e_1, ..., e_n$ and denote by "," the covariant derivation with respect to the Calabi metric $\mathfrak{H}$. The Ricci identity and the Codazzi equations (3.3.3) give

$$\begin{aligned} \Delta A_{ijk} &= \sum A_{ijk,ll} = \sum A_{ijl,kl} \\ &= \sum A_{ijl,lk} + \sum A_{ijr} R_{rlkl} + \sum A_{irl} R_{rjkl} + \sum A_{rjl} R_{rikl} \\ &= \varphi_{,ijk} + \sum A_{ijr} R_{rlkl} + \sum A_{irl} R_{rjkl} + \sum A_{rjl} R_{rikl}. \end{aligned} \tag{4.6.1}$$

Therefore

$$\begin{aligned} \tfrac{n(n-1)}{2} \Delta J &= \tfrac{1}{2} \Delta \left(\sum A_{ijk}^2 \right) = \sum (A_{ijk,l})^2 + \sum A_{ijk} A_{ijk,ll} \\ &= \sum A_{ijk}\, \varphi_{,ijk} + \sum (A_{ijk,l})^2 + \sum A_{ijk} A_{ijr} R_{rlkl} \\ &\quad + \sum (A_{ijk} A_{irl} - A_{ijl} A_{irk}) R_{rjkl}. \end{aligned} \tag{4.6.2}$$

For any $p \in \Omega$, by an appropriate coordinate transformation, we may assume that $f_{ij}(p) = \delta_{ij}$ and $R_{ij}(p) = 0$ for $i \neq j$. From (3.3.4) we have

$$\begin{aligned} R_{ii}(p) &= \tfrac{1}{4}\sum_{m,j}(f_{mij})^2 + \tfrac{n+2}{4}\sum_m f_{mii}\,\tfrac{\partial}{\partial x^m}\ln\rho \\ &\geq \tfrac{1}{4}\left(\sum_m (f_{mii})^2 + (n+2)\sum_m f_{mii}\,\tfrac{\partial}{\partial x^m}\ln\rho\right) \\ &\geq -\tfrac{(n+2)^2}{16}\Phi. \end{aligned} \tag{4.6.3}$$

Inserting (3.3.2) (4.6.3) into (4.6.2) we get

$$\begin{aligned} \tfrac{n(n-1)}{2}\Delta J &= \sum A_{ijk}\varphi_{,ijk} + \sum (A_{ijk,l})^2 + \sum (R_{ijkl})^2 + \sum A_{ijk}A_{ijr}R_{rk} \\ &\geq \sum A_{ijk}\varphi_{,ijk} + \|\nabla A\|^2 + \sum (R_{ijkl})^2 - \tfrac{(n+2)^2}{16}\left(\sum A_{ijk}^2\right)\Phi. \end{aligned} \tag{4.6.4}$$

By (1.1.3), (3.3.5) and the inequality of Schwarz we have

$$\begin{aligned} \sum (R_{ijkl})^2 &\geq \tfrac{2}{n(n-1)}\left(\sum A_{ijk}^2 - \sum (\varphi_i)^2\right)^2 \\ &\geq \tfrac{3}{2n(n-1)}\left(\sum A_{ijk}^2\right)^2 - \tfrac{3(n+2)^4}{8n(n-1)}\Phi^2. \end{aligned} \tag{4.6.5}$$

We substitute (4.6.5) into (4.6.4) and use the inequality of Schwarz; we finally get

$$\Delta J \geq \tfrac{2}{n(n-1)}\sum A_{ijk}\varphi_{,ijk} + \tfrac{2}{n(n-1)}\|\nabla A\|^2 + 2J^2 - \tfrac{(n+2)^4}{4}\Phi^2. \qquad \blacksquare$$

In particular, if f satisfies the PDE (4.5.2), we choose the coordinates $(x^1, ..., x^n)$ at p such that $f_{ij}(p) = \delta_{ij}$, then we have

$$\varphi_{,ijk} = \tfrac{n+2}{2}\left((\ln\rho)_{,l}A_{ijl}\right)_{,k} = \tfrac{n+2}{2}\left((\ln\rho)_{,m}A_{klm}A_{ijl} + (\ln\rho)_{,l}A_{ijl,k}\right).$$

Young's inequality [38] and the inequality of Schwarz give

$$\tfrac{n+2}{n(n-1)}\sum A_{ijk}A_{ijl}A_{klm}(\ln\rho)_{,m} \leq \tfrac{1}{2}J^2 + 8n^2(n-1)^2(n+2)^4\Phi^2, \tag{4.6.6}$$

$$\begin{aligned} \tfrac{n+2}{n(n-1)}\sum A_{ijk}A_{ijl,k}(\ln\rho)_{,l} &\leq \tfrac{2}{n(n-1)}\sum (A_{ijk,l})^2 + \tfrac{(n+2)^2}{8}J\Phi \\ &\leq \tfrac{2}{n(n-1)}\sum (A_{ijk,l})^2 + \tfrac{1}{2}J^2 + \tfrac{(n+2)^4}{128}\Phi^2. \end{aligned} \tag{4.6.7}$$

We insert (4.6.6) (4.6.7) into Lemma 4.6.3 and obtain the following corollary:

Corollary 4.6.4. *If f satisfies the PDE (4.5.2) then we have*

$$\Delta J \geq J^2 - 10(n+2)^8\Phi^2. \qquad \blacksquare$$

4.6.2 *Proof of Theorem 4.6.2*

It is our aim to prove $\Phi \equiv 0$; from the definitions of Φ and ρ, this relation implies that the α-metric and $\mathfrak{H}$ coincide modulo a positive constant factor, and thus

$$\det(f_{ij}) = const$$

everywhere on M. Then Theorem 4.6.2 follows from Theorem 4.6.1.

Denote by $r(p_0, p)$ the geodesic distance function from $p_0 \in M$ with respect to the metric $G^{(\alpha)}$ from section 3.3.5. For any positive number a, let

$$\bar{B}_a(p_0, G^{(\alpha)}) := \{p \in M \,|\, r(p_0, p) \leq a\}.$$

In the following we derive an estimate for $\frac{\Phi}{\rho^\alpha}$ in a geodesic ball $\bar{B}_a(p_0, G^{(\alpha)})$. Set

$$\mathcal{A} := \max_{\bar{B}_a(p_0, G^{(\alpha)})} \left\{(a^2 - r^2)^2 \tfrac{\Phi}{\rho^\alpha}\right\}, \qquad \mathcal{B} := \max_{\bar{B}_a(p_0, G^{(\alpha)})} \left\{(a^2 - r^2)^2 \tfrac{J}{\rho^\alpha}\right\}.$$

Lemma 4.6.5. *Let $f(x^1, ..., x^n)$ be a strictly convex C^∞-function defined on $\bar{B}_a(p_0, G^{(\alpha)})$. If f satisfies the PDE (4.5.2) and $\alpha \neq n+2$ then there exists a constant $C > 0$, depending only on n, α, such that*

$$\mathcal{A} \leq Ca^2, \quad \mathcal{B} \leq Ca^2.$$

Proof. Step 1. We will show that

$$\mathcal{A} \leq C_7 \mathcal{B}^{\frac{1}{2}} a + C_8 a^2,$$

where C_7, C_8 are positive constants depending only on α, n. In analogy to section 4.4 where we used the Blaschke metric, consider the function

$$F := (a^2 - r^2)^2 \tfrac{\Phi}{\rho^\alpha},$$

defined on $\bar{B}_a(p_0, G^{(\alpha)})$. Obviously, F attains its supremum at some interior point p^*. We may assume that r^2 is a C^2-function in a neighborhood of p^*, and $\Phi(p^*) > 0$. Choose an orthonormal frame field on M around p^* with respect to the Calabi metric $\mathfrak{H}$. Then, at p^*,

$$\frac{\Phi_{,i}}{\Phi} - \frac{4rr_{,i}}{a^2 - r^2} - \alpha \frac{\rho_{,i}}{\rho} = 0, \tag{4.6.8}$$

$$\frac{\Delta\Phi}{\Phi} - \frac{\sum(\Phi_{,i})^2}{\Phi^2} - \frac{(n+2)\alpha}{2}\Phi - \frac{8r^2\rho^\alpha}{(a^2 - r^2)^2} - \frac{4r\Delta r}{a^2 - r^2} - \frac{4\rho^\alpha}{a^2 - r^2} \leq 0, \tag{4.6.9}$$

where we use the fact

$$\Delta\rho = \frac{n+4}{2} \frac{\|\text{grad } \rho\|^2}{\rho}, \qquad \|\text{ grad } r \,\|^2 = \rho^\alpha,$$

"," denotes the covariant derivative with respect to the Calabi metric $\mathfrak{H}$ as before. We insert Proposition 4.5.3 into (4.6.9) and get

$$\frac{1}{n-1} \frac{\|\text{grad } \Phi\|^2}{\Phi^2} + \frac{n^2 - 3n - 10}{2(n-1)} \sum \frac{\Phi_{,i}}{\Phi} \frac{\rho_{,i}}{\rho} + \left(\frac{(n+2)^2}{n-1} - \frac{(n+2)\alpha}{2}\right)\Phi - \frac{12a^2\rho^\alpha}{(a^2 - r^2)^2} - \frac{4r\Delta r}{a^2 - r^2} \leq 0.$$

Substituting (4.6.8) and using the inequality of Schwarz yields

$$\frac{(\alpha - (n+2))^2 - \epsilon}{n-1}\Phi - C_1 \frac{a^2\rho^\alpha}{(a^2 - r^2)^2} - \frac{4r\Delta r}{a^2 - r^2} \leq 0, \tag{4.6.10}$$

where ϵ is a small positive constant to be determined later, and C_1 is a positive constant depending only on n, α and ϵ.

Now we calculate the term $\frac{4r\Delta r}{a^2-r^2}$. Denote $a^* = r(p_0, p^*)$. Assume that

$$\max_{\bar{B}_{a^*}(p_0, G^{(\alpha)})} \left\{\frac{\Phi}{\rho^\alpha}\right\} = \frac{\Phi}{\rho^\alpha}(\tilde{p}), \qquad \max_{\bar{B}_{a^*}(p_0, G^{(\alpha)})} \left\{\frac{J}{\rho^\alpha}\right\} = \frac{J}{\rho^\alpha}(q).$$

From the PDE (4.5.2) and the definition of the α-Ricci curvature in (3.3.10), we have

$$R_{ij}^{(\alpha)} = R_{ij} - \frac{(n-2)\alpha}{2}(\ln\rho)_{,k} A_{ijk} + \frac{(n-2)\alpha^2}{4}(\ln\rho)_{,i}(\ln\rho)_{,j} - \frac{1}{2}\left(\frac{(n+2)\alpha}{2} + \frac{(n-2)\alpha^2}{2}\right)\Phi\mathfrak{H}_{ij}.$$

For any $p \in \bar{B}_{a^*}(p_0, G^{(\alpha)})$, by an appropriate coordinate transformation, we may assume that $f_{ij}(p) = \delta_{ij}$ and $R_{ij}(p) = 0$ for $i \neq j$. Then from (4.6.3), using the inequality of Schwarz, we know that the Ricci curvature $Ric(M, G^{(\alpha)})$ with respect to the α-metric on $\bar{B}_{a^*}(p_0, G^{(\alpha)})$ is bounded from below by

$$Ric(M, G^{(\alpha)}) \geq -C_2\left(\frac{\Phi}{\rho^\alpha}(\tilde{p}) + \frac{J}{\rho^\alpha}(q)\right) \tag{4.6.11}$$

for some positive constant C_2, depending only on n, α. By the Laplacian Comparison Theorem (see section 1.3), we get

$$r\Delta^{(\alpha)} r \leq (n-1)\left(1 + C_2\left(\sqrt{\frac{\Phi}{\rho^\alpha}(\tilde{p})} + \sqrt{\frac{J}{\rho^\alpha}(q)}\right) r\right).$$

Thus, using the expression for the Laplacian $\Delta^{(\alpha)}$ in (3.3.9), we obtain

$$\begin{aligned}\frac{4r\Delta r}{a^2-r^2}(p^*) &= \frac{4\rho^\alpha r\Delta^{(\alpha)} r}{a^2-r^2}(p^*) - \frac{2(n-2)\alpha r}{a^2-r^2}\,\mathfrak{H}(\operatorname{grad}\ln\rho, \operatorname{grad} r)(p^*)\\ &\leq \frac{4(n-1)aC_2\rho^\alpha}{a^2-r^2}(p^*)\left(\sqrt{\frac{\Phi}{\rho^\alpha}(\tilde{p})} + \sqrt{\frac{J}{\rho^\alpha}(q)}\right)\\ &\quad + \frac{\epsilon}{n-1}\,\Phi(p^*) + C_3\,\frac{a^2\rho^\alpha}{(a^2-r^2)^2}(p^*)\end{aligned} \tag{4.6.12}$$

for some positive constant C_3 depending only on n, ϵ, α.
We insert (4.6.12) into (4.6.10), this gives

$$\frac{(\alpha-(n+2))^2-2\epsilon}{n-1}\,\frac{\Phi}{\rho^\alpha} \leq \frac{C_4 a^2}{(a^2-r^2)^2} + \frac{4(n-1)aC_2}{a^2-r^2}\left(\sqrt{\frac{\Phi}{\rho^\alpha}(\tilde{p})} + C_3\sqrt{\frac{J}{\rho^\alpha}(q)}\right), \tag{4.6.13}$$

where C_4 is a positive constant depending only on n, ϵ, α.

Step 2. To derive an upper bound for $\frac{\Phi}{\rho^\alpha}$ from (4.6.13), note that

$$\mathcal{A} \geq (a^2-r^2)^2(\tilde{p})\frac{\Phi}{\rho^\alpha}(\tilde{p}) \geq (a^2-r^2)^2(p^*)\frac{\Phi}{\rho^\alpha}(\tilde{p}),$$
$$\mathcal{B} \geq (a^2-r^2)^2(q)\frac{J}{\rho^\alpha}(q) \geq (a^2-r^2)^2(p^*)\frac{J}{\rho^\alpha}(q).$$

Multiplying both sides of (4.6.13) by $(a^2-r^2)^2(p^*)$, we have

$$\frac{(\alpha-(n+2))^2-3\epsilon}{n-1}\mathcal{A} \leq C_5\mathcal{B}^{\frac{1}{2}}a + C_6a^2, \tag{4.6.14}$$

for some positive constants C_5, C_6 depending only on ϵ, α, n.
In case $\alpha \neq n+2$, we may choose ϵ small enough such that $\frac{(\alpha-(n+2))^2-3\epsilon}{n-1} > 0$. Then

$$\mathcal{A} \leq C_7\mathcal{B}^{\frac{1}{2}}a + C_8a^2, \tag{4.6.15}$$

where C_7, C_8 are positive constants depending only on α, n.

Step 3. We will prove the inequality

$$\mathcal{B} \leq C_{13}\mathcal{A} + C_{14}a^2$$

for some positive constants C_{13}, C_{14} depending only on n, α.

Consider the following function

$$\tilde{F} := (a^2 - r^2)^2 \tfrac{J}{\rho^\alpha}$$

defined on $\bar{B}_a(p_0, G^{(\alpha)})$. Obviously, $\tilde{F}$ attains its supremum at some interior point q^*. Then, at q^*, we get

$$\tfrac{J_{,i}}{J} - \tfrac{4rr_{,i}}{a^2-r^2} - \alpha \tfrac{\rho_{,i}}{\rho} = 0, \tag{4.6.16}$$

$$\tfrac{\Delta J}{J} - \tfrac{\sum (J_{,i})^2}{J^2} - \tfrac{(n+2)\alpha}{2}\Phi - \tfrac{8r^2\rho^\alpha}{(a^2-r^2)^2} - \tfrac{4r\Delta r}{a^2-r^2} - \tfrac{4\rho^\alpha}{a^2-r^2} \leq 0. \tag{4.6.17}$$

We discuss two cases:

Case 1. $\frac{J}{\rho^\alpha}(q^*) \leq \frac{\Phi}{\rho^\alpha}(q^*)$, then

$$\mathcal{B} = (a^2 - r^2)^2 \tfrac{J}{\rho^\alpha}(q^*) \leq (a^2 - r^2)^2 \tfrac{\Phi}{\rho^\alpha}(q^*) \leq \mathcal{A}.$$

Then Step 3 is complete.

Case 2. $\frac{J}{\rho^\alpha}(q^*) > \frac{\Phi}{\rho^\alpha}(q^*)$. We use Corollary 4.6.4, the inequality $1 > \frac{\Phi}{J}(q^*)$, (4.6.16) and the inequality of Schwarz to obtain

$$J - C_9\Phi - \tfrac{44a^2\rho^\alpha}{(a^2-r^2)^2} - \tfrac{4r\Delta r}{a^2-r^2} \leq 0, \tag{4.6.18}$$

where C_9 is a positive constant depending only on n, α.

Denote $b^* := r(p_0, q^*)$. Assume that

$$\max_{\bar{B}_{b^*}(p_0, G^{(\alpha)})} \left\{\tfrac{\Phi}{\rho^\alpha}\right\} = \tfrac{\Phi}{\rho^\alpha}(\tilde{p}_1), \qquad \max_{\bar{B}_{b^*}(p_0, G^{(\alpha)})} \left\{\tfrac{J}{\rho^\alpha}\right\} = \tfrac{J}{\rho^\alpha}(q_1).$$

As in step 1, from the inequality of Schwarz and Young's inequality (see [38]) we get, at q^*,

$$\tfrac{4r\Delta r}{a^2-r^2} \leq \tfrac{4(n-1)aC_2\rho^\alpha}{a^2-r^2}\left(\sqrt{\tfrac{\Phi}{\rho^\alpha}(\tilde{p}_1)} + \sqrt{\tfrac{J}{\rho^\alpha}(q_1)}\right) + \Phi + C_{10}\, \tfrac{a^2\rho^\alpha}{(a^2-r^2)^2}, \tag{4.6.19}$$

for some positive constant C_{10} depending only on n, α.
We insert (4.6.19) into (4.6.18) and use the inequality of Schwarz again, then:

$$\begin{aligned}(a^2 - r^2)^2 \tfrac{J}{\rho^\alpha}(q^*) \leq C_{11}(a^2 - r^2)^2 \tfrac{\Phi}{\rho^\alpha}(q^*) + (a^2 - r^2)^2(q^*)\tfrac{\Phi}{\rho^\alpha}(\tilde{p}_1)\\ + \tfrac{1}{4}(a^2 - r^2)^2(q^*)\tfrac{J}{\rho^\alpha}(q_1) + C_{12}a^2,\end{aligned}$$

where C_{11}, C_{12} are positive constants depending only on n, α.
Using the same method as in Step 1, we obtain the inequality

$$\mathcal{B} \leq C_{13}\mathcal{A} + C_{14}a^2 \tag{4.6.20}$$

for some positive constants C_{13}, C_{14} depending only on n, α. Step 3 is complete.

Step 4. From (4.6.20) and (4.6.15), there exits a positive constant C depending only on n, α such that

$$\mathcal{A} \leq Ca^2, \quad \mathcal{B} \leq Ca^2.$$

■

Proof of Theorem 4.6.2. Using Lemma 4.6.5, at any interior point of $B_a(p_0, G^{(\alpha)})$, we obtain

$$\tfrac{\Phi}{\rho^\alpha} \leq C \tfrac{a^2}{(a^2-r^2)^2}.$$

For $a \to \infty$ we get

$$\Phi \equiv 0.$$

This means that M is an affine complete parabolic affine hypersphere. We apply Theorem 4.6.1 and conclude that M must be an elliptic paraboloid. This completes the proof of Theorem 4.6.2. ■

Comment on the proof. Recall the comparison of two geometric proofs from section 4.5.5. In a similar way we would like to comment on the proof of Theorem 4.6.2.
Again we use the Calabi normalization and the Calabi metric for the geometric modelling; we aim to show that $\Phi \equiv 0$ on M, that means to show that the Tchebychev vector field satisfies $T \equiv 0$ (compare the comment on the extension of the Theorem of Jörgens-Calabi-Pogorelov for $n \leq 4$ in section 4.5.5); this way we aim to reduce the problem to Calabi's Theorem 4.6.1.
As before we apply two differential inequalities.
(i) The first one is given in Corollary 4.6.4; it is an analogue of (4.1.7).
(ii) For a second differential inequality (4.6.9) we consider the functions

$$F := (a^2 - r^2)^2 \tfrac{\Phi}{\rho^\alpha} \qquad \text{and} \qquad \tilde{F} := (a^2 - r^2)^2 \tfrac{J}{\rho^\alpha}$$

on the geodesic ball $B_a(p_0, G^{(\alpha)})$ with respect to the metric $G^{(\alpha)}$, and apply the Laplacian Comparison Theorem again to get (4.6.13) (note: according to the assumptions the $\alpha-$metric is complete).
To get an upper estimate for Φ on the geodesic ball we need the joint estimate for the maxima $\mathcal{A}$ and $\mathcal{B}$ in Lemma 4.6.5; this need follows from (4.6.15) and (4.6.20). For any interior point $p \in B_a(p_0, G^{(\alpha)})$ this finally leads to the estimate

$$\tfrac{\Phi}{\rho^\alpha}(p) \leq C \cdot \tfrac{1}{a^2(1-\frac{r^2}{a^2})^2}.$$

We apply the completeness assumption: then $\Phi(p) \to 0$ for $a \to \infty$.

Chapter 5

Affine Maximal Hypersurfaces

5.1 The First Variation of the Equiaffine Volume Functional

We consider a non-degenerate hypersurface $x : M \to \mathbb{R}^{n+1}$ in unimodular affine space and a (sufficiently small) domain $D \subset M$ with boundary ∂D; its *volume* is

$$V(D) = \int_D dV, \tag{5.1.1}$$

where the equiaffine volume form was calculated in section 2.2.1. We wish to compute the first variation $\delta V(D)$ of V, keeping the boundary ∂D fixed, cf. [27].

We describe this situation analytically. For this, let I be the open interval $-\frac{1}{2} < t < \frac{1}{2}$ and $f : M \times I \to \mathbb{R}^{n+1}$ be a smooth mapping such that its restriction to $M \times \{t\}$, for any $t \in I$, is an immersion, and where $f(p, 0) = x(p)$ for $p \in M$. We consider a frame field $e_\alpha(p,t)$ over $M \times I$ such that, for every $t \in I$, $e_i(p,t)$ are tangent vectors, and $e_{n+1}(p,t)$ is parallel to the affine normal of the immersion $f(M \times t)$ at (p,t). We pull the forms $\omega^\alpha, \omega_\alpha^\beta$ in the frame manifold back to $M \times I$; since the vectors e_i span the tangent hyperplane at $f(p,t)$, we have

$$\omega^{n+1} = adt. \tag{5.1.2}$$

Its exterior differentiation gives

$$\sum \omega^i \wedge \omega_i^{n+1} + dt \wedge \left(a\, \omega_{n+1}^{n+1} + da\right) = 0. \tag{5.1.3}$$

Thus we can set

$$\begin{aligned} \omega_i^{n+1} &= \sum h_{ij}\omega^j + h_i dt, \\ a\omega_{n+1}^{n+1} + da &= \sum h_i\omega^i + hdt, \end{aligned} \tag{5.1.4}$$

where, as before,

$$h_{ij} = h_{ji}.$$

Exterior differentiation of (5.1.4) gives

$$\begin{aligned} &\sum \left(dh_{ij} - \sum h_{ik}\, \omega_j^k - \sum h_{jk}\, \omega_i^k + h_{ij}\, \omega_{n+1}^{n+1}\right) \wedge \omega^j \\ &+ \sum \left(dh_i - \sum h_k\, \omega_i^k + h_i\, \omega_{n+1}^{n+1} - a \sum h_{ij}\, \omega_{n+1}^j\right) \wedge dt = 0. \end{aligned} \tag{5.1.5}$$

Hence we can write (see section 2.4.1):

$$dh_{ij} = \sum h_{ik}\,\omega_j^k + \sum h_{jk}\,\omega_i^k - h_{ij}\,\omega_{n+1}^{n+1} + \sum h_{ijk}\,\omega^k + p_{ij}dt, \tag{5.1.6}$$
$$dh_i = \sum h_k\,\omega_i^k \; - \; h_i\,\omega_{n+1}^{n+1} + a\sum h_{ij}\,\omega_{n+1}^j + \sum p_{ij}\,\omega^j + q_i\,dt,$$

where h_{ijk} is symmetric in all three indices, and

$$p_{ij} = p_{ji}.$$

Let (H^{ik}) be the adjoint matrix of (h_{ik}), so that, with $H := \det(h_{ik})$,

$$\sum H^{ij}\,h_{jk} = \delta_k^i\,H. \tag{5.1.7}$$

By (5.1.6) we have

$$dH = \sum H^{ij}\,dh_{ij} = -\,(n+2)\,H\,\omega_{n+1}^{n+1} + \sum H^{ij}\,h_{ijk}\,\omega^k + \sum H^{ij}\,p_{ij}\,dt. \tag{5.1.8}$$

An appropriate change of frames

$$e_i^* = \sum a_i^k e_k, \qquad e_{n+1}^* = A^{-1}e_{n+1} + \sum a_{n+1}^i e_i\,,$$

where $A = \det\left(a_i^k\right)$, gives

$$\sum H^{ij}\,h_{ijk} = 0.$$

Geometrically this means that e_{n+1} is parallel to the affine normal of the hypersurface $f\,(M\times t)$ at $f\,(p,t)$. Now the resulting equation (5.1.8) can be written as

$$f^*\omega_{n+1}^{n+1} + \tfrac{1}{n+2}\,d\ln|H| = b\,dt, \tag{5.1.9}$$

where

$$b := \tfrac{1}{(n+2)H}\sum H^{ij}\,p_{ij}.$$

For later application we differentiate (5.1.7) and use (5.1.6) to obtain

$$dH^{ij} = \sum -H^{ik}\,\omega_k^j - \sum H^{jk}\,\omega_k^i + H^{ij}\left(\omega_{n+1}^{n+1} + \; d\ln|H|\right)$$
$$- \tfrac{1}{H}\sum H^{ik}\,H^{jl}\,h_{klr}\,\omega^r - \tfrac{1}{H}\sum H^{ik}\,H^{jl}\,p_{kl}\,dt. \tag{5.1.10}$$

We abbreviate

$$h^i := \sum H^{ij}h_j.$$

(5.1.6) and (5.1.10) imply

$$dh^i = \sum -h^k\,\omega_k^i + h^i\,d\ln|H| + aH\omega_{n+1}^i \; - \; \tfrac{1}{H}\sum H^{ik}\,h^l\,h_{klr}\,\omega^r$$
$$+ \sum H^{ij}\,p_{jk}\,\omega^k - \tfrac{1}{H}\sum H^{ik}\,h^l\,p_{kl}\,dt + \sum H^{ij}\,q_j\,dt. \tag{5.1.11}$$

We use (2.1.5) and the Maurer-Cartan equations (2.1.6) and obtain

$$\begin{aligned} & d\left(\omega^1 \wedge \cdots \wedge \omega^n\right) \\ &= \sum_i (-1)^{i-1} \omega^1 \wedge \cdots \wedge \omega^{i-1} \wedge \left(\omega^i \wedge \omega_i^i + \omega^{n+1} \wedge \omega_{n+1}^i\right) \wedge \omega^{i+1} \wedge \cdots \wedge \omega^n \\ &= \omega_{n+1}^{n+1} \wedge \omega^1 \wedge \cdots \wedge \omega^n \\ &\qquad + \omega^{n+1} \wedge \sum_i \omega^1 \wedge \cdots \wedge \omega^{i-1} \wedge \omega_{n+1}^i \wedge \omega^{i+1} \wedge \cdots \wedge \omega^n. \end{aligned} \tag{5.1.12}$$

Pulling back under f, we get

$$\begin{aligned} & |H|^{-\frac{1}{n+2}} d\left(|H|^{\frac{1}{n+2}} f^*\left(\omega^1 \wedge \cdots \wedge \omega^n\right)\right) \\ &\qquad = \left(f^*\omega_{n+1}^{n+1} + \tfrac{1}{n+2} d \ln |H| - n\,|H|^{-\frac{1}{n+2}} L_1 a dt\right) \wedge f^*\left(\omega^1 \wedge \cdots \wedge \omega^n\right) \\ &\qquad = \left(b - n\,|H|^{-\frac{1}{n+2}} L_1 a\right) dt \wedge \omega^{*1} \wedge \cdots \wedge \omega^{*n}, \end{aligned}$$

where ω^{*i} is defined from $f^*\omega^i$ by "splitting off" the term in dt:

$$f^*\omega^i = \omega^{*i} + a^i dt. \tag{5.1.13}$$

Analogously we decompose the operator d on $M \times I$:

$$d_M + dt \tfrac{\partial}{\partial t}. \tag{5.1.14}$$

In the above equation we equate the terms in dt and get

$$\begin{aligned} & \tfrac{\partial}{\partial t}\left(|H|^{\frac{1}{n+2}} \omega^{*1} \wedge \cdots \wedge \omega^{*n}\right) \\ &\qquad + d_M \left\{ |H|^{\frac{1}{n+2}} \sum_i (-1)^i a^i \omega^{*1} \wedge \cdots \wedge \omega^{*,i-1} \wedge \omega^{*,i+1} \wedge \cdots \wedge \omega^{*n} \right\} \\ &= |H|^{\frac{1}{n+2}} \left(b - n\,|H|^{-\frac{1}{n+2}} L_1\, a\right) \omega^{*1} \wedge \cdots \wedge \omega^{*n}. \end{aligned}$$

On ∂D we have $a^i = 0$. Integrating over D and setting $t = 0$, we find the first variation of the volume

$$\begin{aligned} V'(0) &= \frac{\partial}{\partial t} \int_D |H|^{\frac{1}{n+2}} \omega^{*1} \wedge \cdots \wedge \omega^{*n} \,|_{t=0} \\ &= \int_D \left(b - n\,|H|^{-\frac{1}{n+2}} L_1 a\right) dV \,|_{t=0}\,. \end{aligned} \tag{5.1.15}$$

The last expression can be simplified; we prove the following:

Lemma 5.1.1. *For $t = const$, the form*

$$\left(b - \tfrac{n}{n+2} |H|^{-\frac{1}{n+2}} L_1 a\right) dV = \left(b\,|H|^{\frac{1}{n+2}} - \tfrac{n}{n+2} L_1 a\right) \omega^1 \wedge \cdots \wedge \omega^n$$

is exact and its integral over D, for $t = 0$, is zero.

Proof. We introduce the form

$$\Omega := \tfrac{1}{(n-1)!} \sum_{\epsilon_{i_1 \cdots i_n}} h^{i_1} \omega^{i_2} \wedge \cdots \wedge \omega^{i_n}$$
$$= \sum_i (-1)^{i-1} \omega^1 \wedge \cdots \wedge \omega^{i-1} h^i \wedge \omega^{i+1} \wedge \cdots \wedge \omega^n, \tag{5.1.16}$$

where $\epsilon_{i_1 \cdots i_n}$ is $+1$ or -1 if $\{i_1, \cdots, i_n\}$ is an even or odd permutation of $\{1, 2, \cdots, n\}$, and otherwise is zero. From (2.1.5) and (5.1.11) we find

$$d\Omega = \left(\omega_{n+1}^{n+1} + \tfrac{dH}{H}\right) \wedge \Omega + \left\{-a \sum l_i^i + (n+2)\, b\right\} H \omega^1 \wedge \cdots \wedge \omega^n. \tag{5.1.17}$$

It follows that

$$d\left(|H|^{-\frac{n+1}{n+2}} \Omega\right) = |H|^{-\frac{n+1}{n+2}} \left(d\Omega - \tfrac{n+1}{n+2}\, d\ln|H| \wedge \Omega\right).$$

By (5.1.9) we have, for $t = const$,

$$d\left(|H|^{-\frac{n+1}{n+2}} \Omega\right) = (\text{sgn } H)\left\{-a \sum l_i^i + (n+2)\, b\right\} |H|^{\frac{1}{n+2}} \omega^1 \wedge \cdots \wedge \omega^n. \tag{5.1.18}$$

Here sgn denotes the sign.
For the variation, we assume that $h_i(p, 0) = 0$ for $p \in \partial D$. Hence the lemma follows. ■

Together with (5.1.15) we arrive at the following result.

Proposition 5.1.2.

$$V'(0) = -\tfrac{n(n+1)}{n+2} \int_D |H|^{-\frac{1}{n+2}} L_1\, a\, dV \,|_{t=0} . \tag{5.1.19}$$

Corollary 5.1.3. *On a locally strongly convex hypersurface, if* $V'(0) = 0$ *for an arbitrary function* $a : D \times I \to \mathbb{R}$, *satisfying*

$$a(p, 0) = 0, \quad h_i(p, 0) = 0, \quad p \in \partial D,$$

we must have $L_1 = 0$, *i.e.,* M *is an affine extremal hypersurface.*

5.2 Affine Maximal Hypersurfaces

5.2.1 *Graph hypersurfaces*

Let $x : M \to \mathbb{R}^{n+1}$ be a locally strongly convex hypersurface; the parameter manifold M may be open or compact, and if it has a boundary ∂M this should be smooth.

Definition 5.2.1. **(a)** An *allowable interior deformation* of x is a differentiable map $f : M \times I \to \mathbb{R}^{n+1}$, where $I := (-\epsilon, \epsilon)$ with $\epsilon > 0$ is an open interval such that f has the following properties:
(i) For each $t \in I$ the map $x_t : M \to \mathbb{R}^{n+1}$, defined by $x_t(p) = f(p, t)$, is a locally

strongly convex hypersurface, and $x_0 = x$ for $t = 0$.
(ii) There exists a compact subdomain $\bar{N} \subset M$, where $\bar{N}$ is the closure of a connected, open subset of $N \subset M$ with smooth boundary ∂N, and where ∂N may contain, meet, or be disjoint from ∂M, such that, for each $p \in M \setminus \bar{N}$ and all $t \in I$, $f(p,t) = x(p)$.
(iii) For each $p \in \partial\bar{N}$ and for all $t \in I$, $f(p,t) = x(p)$, the tangent hyperplane $dx_t(p)$ coincides with $dx(p)$.
(b) A locally strongly convex hypersurface $x^\sharp : M \to \mathbb{R}^{n+1}$ is said to be *interior-homotopic* to x, if there exists an allowable interior deformation $f : M \times I \to \mathbb{R}^{n+1}$ with $I = (-\epsilon, 1+\epsilon)$ such that $x_0 = x$, $x_1 = x^\sharp$.

In the sequel, when we study variations of the affine invariant volume of $x(M)$ under interior deformations, we may replace M, without loss of generality, by the compact subdomain $\bar{N} \subset M$, or, from the beginning, simply assume that M is compact with smooth boundary.

Definition 5.2.2. Let $x : M \to \mathbb{R}^{n+1}$ be a locally strongly convex hypersurface. If $L_1 = 0$ on M then $x(M)$ is called an *affine maximal hypersurface.*

It is a consequence from (5.1.19) that affine maximal hypersurfaces are critical points of the equiaffine volume functional. Considering the analogy that in the Euclidean and in the affine hypersurface theory both Euler-Lagrange equations are given by the vanishing of the trace of the associated shape operator, Blaschke and his school originally called hypersurfaces with $L_1 = 0$ *affine minimal hypersurfaces* without calculating the second variation of the volume functional. 60 years later this was done by Calabi [21], and he suggested to call locally strongly convex hypersurfaces with $L_1 = 0$ *affine maximal hypersurfaces* because of the following result (in fact, Calabi's result is a little more general than the following Theorem 5.2.3, see [21]).

Theorem 5.2.3. *Let x, $x^\sharp : \Omega \to \mathbb{R}^{n+1}$ be two graphs, defined on a compact domain by locally strongly convex functions $f, f^\sharp$, namely*

$$x^{n+1} = f(x) \quad \text{and} \quad x^{n+1} = f^\sharp(x), \quad \text{where} \quad x = \left(x^1, \cdots, x^n\right)$$

resp.; we use an obvious notation to denote invariants of $f^\sharp$. Assume that, at the boundary $\partial\Omega$, we have the relations $f = f^\sharp$ and $\frac{\partial f}{\partial x^i} = \frac{\partial f^\sharp}{\partial x^i}$ for $i = 1, 2, \cdots, n$. If $L_1 = 0$ on Ω then

$$\int_\Omega dV \geq \int_\Omega dV^\sharp,$$

and equality holds if and only if $f = f^\sharp$ on Ω.

Proof. Choose an allowable interior deformation defined by the linear interpolation $x_t : f_t(x) = (1-t) f(x) + t f^\sharp(x) = f(x) + t\left(f^\sharp(x) - f(x)\right), \quad x \in \Omega, \quad 0 \leq t \leq 1.$ Then $f_t(x)$ is locally strongly convex everywhere in Ω. Let $\mu_1(x), \cdots, \mu_n(x)$, for each $x \in \Omega$, denote the eigenvalues of the matrix

$$C^i_j := \sum f^{ik}\left(f^\sharp_{kj} - f_{kj}\right),$$

where we use the local notation from sections 1.1.1 and 3.3.4. For each t, the eigenvalues of the matrix

$$\left(\sum_{k=1}^{n} f^{ik} \frac{\partial^2 f_t}{\partial x^k \partial x^j}\right)$$

are positive, namely they are given by

$$1 + t\mu_i(x), \qquad 1 \le i \le n.$$

The volume element dV_t of x_t satisfies

$$dV_t = \left[\det\left(\frac{\partial^2 f_t}{\partial x^j \partial x^i}\right)\right]^{\frac{1}{n+2}} dx^1 \wedge \cdots \wedge dx^n = \prod_{i=1}^{n} (1 + t\mu_i(x))^{\frac{1}{n+2}} dV.$$

We apply the well known geometric-arithmetic-mean inequality to the $(n+2)$ positive numbers $1,\ 1,\ 1 + t\mu_1,\ \ldots,\ 1 + t\mu_n$ and get

$$dV_t \le \frac{1}{n+2}\left(2 + \sum_{i=1}^{n} (1 + t\mu_i(x))\right) dV = \left[1 + \frac{t}{n+2}\left(\sum_{i=1}^{n} \mu_i(x)\right)\right] dV;$$

equality holds if and only if either $t = 0$ or $\mu_1(x) = \mu_2(x) = \cdots = \mu_n(x) = 0$. Then

$$\frac{\partial (dV_t(x))}{\partial t}|_{t=0} = \frac{1}{n+2}\left(\sum_{i=1}^{n} \mu_i(x)\right) dV$$

implies

$$dV_t \le dV + t\,\frac{\partial (dV_t)}{\partial t}\,|_{t=0}\,.$$

Thus

$$\int_\Omega dV_t \le \int_\Omega dV + t\,\frac{\partial}{\partial t}\int_\Omega dV_t\,|_{t=0}\,.$$

From formula (5.1.19) and $L_1 = 0$ we have

$$\frac{\partial}{\partial t}\int_\Omega dV_t\,|_{t=0} = 0,$$

therefore

$$\int_\Omega dV^\sharp \le \int_\Omega dV\,.$$

The equality $\int_\Omega dV^\sharp = \int_\Omega dV$ implies that $\mu_1(x) = \mu_2(x) = \cdots = \mu_n(x) = 0$ for all $x \in \Omega$. This means that $\left(f_{ij} - f^\sharp_{ij}\right)$ is identically zero. As $f = f^\sharp$ on the boundary $\partial\Omega$, we finally have $f = f^\sharp$ on Ω. ■

While the foregoing result is restricted to locally strongly convex graph hypersurfaces in arbitrary dimension, Calabi proved that, for any affine extremal locally strongly convex surface in $\mathbb{R}^3$, the second variation, under all interior deformations of the equiaffinely invariant volume functional, is negative definite (see [21], Theorem 1.3).

5.2.2 *The PDE for affine maximal hypersurfaces*

We derive the differential equation of an affine maximal hypersurface. Again, let $x : \Omega \to \mathbb{R}^{n+1}$ be the graph of a strictly convex function

$$x^{n+1} = f(x^1, \cdots, x^n), \qquad \text{where} \qquad (x^1, \cdots, x^n) \in \Omega \subset \mathbb{R}^n.$$

We choose a unimodular affine frame field as in section 2.7 and recall from there the representation of the Blaschke metric G and the conormal field U. The conormal field satisfies the Schrödinger type PDE (2.3.13), where $L_1 = 0$ for an affine maximal hypersurface. This gives:

Theorem 5.2.4. *Let $x : M \to \mathbb{R}^{n+1}$ be a locally strongly convex hypersurface, given as graph of a function f; x is an affine maximal hypersurface if and only if f satisfies*

$$\Delta \left\{ \left[\det \left(\frac{\partial^2 f}{\partial x^j \partial x^i} \right) \right]^{\frac{-1}{n+2}} \right\} = 0.$$

Remarks. **(a)** In section 1.1.2 there is the local representation of the Laplacian. **(b)** Obviously, any parabolic affine hypersphere is an affine maximal hypersurface. In particular, the elliptic paraboloid

$$x^{n+1} = \tfrac{1}{2} \left[\left(x^1\right)^2 + \cdots + \left(x^n\right)^2 \right], \qquad (x^1, \cdots, x^n) \in \mathbb{R}^n,$$

is an affine-complete, affine maximal hypersurface.

About complete affine maximal surfaces there are two famous conjectures, one is Chern's conjecture (see [26], [27]), the other is called Calabi's conjecture [21].

Chern's conjecture 5.2.5. *Let $x^3 = f(x^1, x^2)$ be a strictly convex function defined for all $\left(x^1, x^2\right) \in \mathbb{R}^2$. If the graph $M = \left\{(x^1, x^2, f(x^1, x^2)) \mid (x^1, x^2) \in \mathbb{R}^2\right\}$ is an affine maximal surface then M must be an elliptic paraboloid.*

Calabi's conjecture 5.2.6. *A locally strongly convex affine complete surface $x : M \to \mathbb{R}^3$ with affine mean curvature $L_1 \equiv 0$ is an elliptic paraboloid.*

These conjectures were generalized to higher dimensions [58]. The two conjectures above differ in the assumption on the completeness of the affine maximal hypersurface considered. Both problems are called an *affine Bernstein problem.* Both problems were long standing open problems.

Remark. We recall different notions of completeness in affine hypersurface theory from section 4.2. In Chern's conjecture one assumes Euclidean completeness, in Calabi's conjecture one assumes affine completeness. In section 4.2 we showed that both completeness assumptions are not equivalent. In 2000, Trudinger and Wang [91] solved Chern's conjecture in dimension $n = 2$. Later, Li and Jia [52], and also Trudinger and Wang [92], solved Calabi's conjecture for two dimensions independently, using quite different methods. Li and Jia used a blow up analysis to show that, for an affine complete maximal surface, $\|B\|_G$ (the tensor norm of the Weingarten tensor B) is bounded above, then they used a result of Martinez and Milan [70] to complete the proof of Calabi's conjecture (for details see section 5.5 below). Trudinger and Wang showed that, for affine maximal surfaces, affine completeness implies Euclidean completeness (see section 4.2.1); then the proof of Calabi's conjecture follows from the proof of Chern's conjecture. So far the higher dimensional affine Bernstein problems are unsolved. In this monograph we first give a proof of Calabi's Conjecture, and then give two different proofs of Chern's Conjecture in two dimensions.

As a first step we state a result of Calabi. Under an additional assumption, he proved the following result [21].

Proposition 5.2.7. *Let $x : M \to \mathbb{R}^3$ be a Euclidean complete affine maximal surface. If M is also affine complete then $x(M)$ is an elliptic paraboloid.*

Proof. Since $x(M)$ is locally strongly convex and Euclidean complete, it follows from the theorem of Hadamard-Sackstedter-Wu (see section 1.2.1) that $x(M)$ is the graph of some positive convex function, say $f(x^1, x^2)$. The normed scalar curvature satisfies the equiaffine Theorema Egregium: $\kappa = J + L_1 = J \geq 0$; f is convex thus $\det\left(\frac{\partial^2 f}{\partial x^j \partial x^i}\right) > 0$; the PDE for an affine maximal surface reads

$$\Delta F := \Delta\left\{\left[\det\left(\frac{\partial^2 f}{\partial x^j \partial x^i}\right)\right]^{-\frac{1}{4}}\right\} = 0.$$

The positive function F is harmonic on a complete Riemannian 2-manifold with non-negative curvature, thus $F = const$ according to a theorem of Yau (see section 1.2.2). From section 4.1.1 we know that the equation $\det\left(\frac{\partial^2 f}{\partial x^j \partial x^i}\right) = const$ leads to an improper affine sphere, and the completeness of (M, G) gives the assertion. ∎

5.3 An Affine Analogue of the Weierstrass Representation

5.3.1 *The representation formula*

Again, we consider a Euclidean inner product on V with normed determinant form Det. We study affine maximal surfaces in $\mathbb{R}^3$.

Let $x : M \to \mathbb{R}^3$ be a locally strongly convex surface. Choose isothermal parameters u, v with respect to the Blaschke metric G, and let $e_1 := \partial_u x = x_u$, $e_2 := \partial_v x = x_v$; denote $U_u := \frac{\partial U}{\partial u}$, $U_v := \frac{\partial U}{\partial v}$. Then $G_{11} = G_{22} > 0, G_{12} = G_{21} = 0$. We have

$$\begin{aligned} &\langle U, x_u\rangle = 0, \quad \langle U_v, x_u\rangle = 0, \quad \langle U_u, x_u\rangle = -G_{11}, \\ &\langle U, x_v\rangle = 0, \quad \langle U_u, x_v\rangle = 0, \quad \langle U_v, x_v\rangle = -G_{22}, \\ &\langle U, Y\rangle = 1, \quad \langle U_u, Y\rangle = 0, \quad \langle U_v, Y\rangle = 0. \end{aligned} \tag{5.3.1}$$

We use the cross product construction from section 2.1.1; then

$$\begin{aligned} x_u &= \lambda[U, U_v], \qquad x_v = \mu[U, U_u], \\ Det(x_u, x_v, Y) &\cdot Det(U_u, U_v, U) = G_{11}^2, \end{aligned} \tag{5.3.2}$$

where λ, μ are differentiable functions. Since

$$Det(x_u, x_v, Y) = |\det(h_{ij})|^{\frac{1}{4}} = G_{11},$$

we have

$$Det(U_u, U_v, U) = G_{11} > 0.$$

From (5.3.1) and (5.3.2),

$$-G_{11} = (U_u, x_u) = -\lambda \cdot Det(U_u, U_v, U) = -\lambda \cdot G_{11}.$$

Hence $\lambda = 1$. Similarly, we have $\mu = -1$. Consequently

$$x_u = [U, U_v], \qquad x_v = -[U, U_u]. \tag{5.3.3}$$

Thus we obtain the following analogue of the Weierstrass representation:

$$x = \int [U, U_v]du \ - \ [U, U_u]dv. \tag{5.3.4}$$

If $x(M)$ is an affine maximal surface, then

$$\Delta U = 0,$$

where, in the given coordinate system, the Laplacian simplifies to:

$$\Delta = \frac{1}{Det(U_u, U_v, U)} \left(\frac{\partial^2}{\partial u^2} + \frac{\partial^2}{\partial v^2} \right).$$

It follows that the components $U^1(u,v)$, $U^2(u,v)$ and $U^3(u,v)$ of U are harmonic functions.

Conversely, consider a given triple of functions

$$U = \left(U^1(u,v),\, U^2(u,v),\, U^3(u,v)\right),$$

defined on a simply connected domain $\Omega \subset \mathbb{R}^2$, that satisfies the following two conditions:

(i) U^1, U^2, U^3 are harmonic with respect to the canonical metric of $\mathbb{R}^2$;
(ii) $Det(U_u, U_v, U) > 0$ in Ω.

Then we can construct an affine maximal surface $x : \Omega \to A^3$ as follows:

$$x(u,v) = \int_{(u_0,v_0)}^{(u,v)} [U, U_v]du \ - \ [U, U_u]dv, \tag{5.3.4$'$}$$

where $(u_0, v_0), (u,v) \in \Omega$. The surface is well defined because the integrability conditions are satisfied:

$$[U, U_v]_v + [U, U_u]_u \ = \ [U, U_{uu} + U_{vv}] = 0.$$

Now let us prove that the surface, defined by (5.3.4)′, is a locally strongly convex affine maximal surface. From (5.3.4)′ we have

$$x_u = [U, U_v], \qquad x_v = -[U, U_u], \tag{5.3.3$'$}$$

$$[x_u, x_v] = Det(U_u, U_v, U) \cdot U. \tag{5.3.5}$$

Define

$$e_3 := \frac{U}{Det(U_u, U_v, U) \cdot \langle U, U \rangle},$$

then

$$Det(x_u, x_v, e_3) = 1,$$

i.e., $\{x; x_u, x_v, e_3\}$ is a unimodular affine frame field, and the structure equations of Gauß read

$$x_{ij} = \sum \Gamma_{ij}^k x_k + h_{ij} e_3, \qquad 1 \le i, j \le 2.$$

From (5.3.3)′ and (5.3.5) we have

$$h_{11} = h_{22} = [Det(U_u, U_v, U)]^2, \qquad h_{12} = h_{21} = 0.$$

Hence the bilinear form h and the Blaschke metric G satisfy

$$G_{ij} = [\det(h_{kl})]^{-\frac{1}{4}} h_{ij},$$

i.e., $\qquad G_{11} = G_{22} = Det(U_u, U_v, U) \qquad$ and $\qquad G_{12} = G_{21} = 0.$

Thus G is positive definite and therefore $x(M)$ is locally strongly convex. The conormal vector is given by

$$[\det(h_{kl})]^{-\frac{1}{4}} \cdot [x_u, x_v] = U.$$

Since $U^1(u,v)$, $U^2(u,v)$ and $U^3(u,v)$ are harmonic functions, $x(M)$ is an affine maximal surface. We summarize the foregoing results:

Affine Weierstrass Representation. *Consider* $\mathbb{R}^3$ *with a Euclidean inner product* $\langle\ ,\ \rangle : \mathbb{R}^3 \times \mathbb{R}^3 \to \mathbb{R}$ *and a smooth map* $U : \Omega \to \mathbb{R}^3$, *where* $\Omega \subset \mathbb{R}^2$ *is a simply connected domain. Define* $x : \Omega \to \mathbb{R}^3$ *by*

$$x(u,v) = \int_{(u_0,v_0)}^{(u,v)} [U, U_v]\, du \; - \; [U, U_u]\, dv.$$

Then x *is an affine maximal surface if and only if* U *with components* U^i $(i = 1,2,3)$ *satisfies the above conditions* **(i)** *and* **(ii)**.

5.3.2 *Examples*

In the following we give some examples of affine maximal surfaces.
Example 1. Consider $\Omega = \mathbb{R}^2$ and define $U : \mathbb{R}^2 \to \mathbb{R}^3$ by $U := (1, u, v)$; then

$$x = \left(\tfrac{1}{2}(u^2 + v^2),\ -u,\ -v\right)$$

is an elliptic paraboloid.
Example 2. Consider $\Omega := \{(u,v) \in \mathbb{R}^2 \mid u > 0\}$ and define $U : \Omega \to \mathbb{R}^3$ by $U := (1, u^2 - v^2, v)$, then

$$Det(U_u, U_v, U) = 2u.$$

The construction above gives

$$x = \left(\tfrac{1}{3}u^3 + uv^2, -u, -2uv\right), \qquad u > 0.$$

Example 3. Let $\Omega := \{(u,v) \in \mathbb{R}^2 \mid u > 0,\ \ v < 0\}$ and define $U := (u, v, 2uv)$, then

$$Det(U_u, U_v, U) = -2uv.$$

The integration of $(5.3.4)'$ gives

$$x = \left(-\tfrac{2}{3}v^3,\ -\tfrac{2}{3}u^3,\ \tfrac{1}{2}(u^2 + v^2)\right), \qquad u > 0,\ v < 0.$$

From the point of view of local differential geometry, the formula (5.3.4) admits the construction of all affine maximal surfaces.

5.4 Calabi's Computation of ΔJ in Holomorphic Terms

This section contains a different and very elegant computation of ΔJ in terms of the Blaschke geometry; it is due to Calabi; he used holomorphic parameters in an elegant special notation. We will follow this notation.
First of all, let us express the structure equations of an affine surface in terms of holomorphic parameters (see [22], [99]).
Let $x : M \to \mathbb{R}^3$ be a locally strongly convex surface. The affine structure of $\mathbb{R}^3$ induces an orientation and a conformal structure on M, namely by suitably oriented relative normalizations (see section 3.1). On $\mathbb{R}^3$, define additionally an appropriate Euclidean structure; then the conformal class contains the (positive definite) Euclidean second fundamental form of x

$$II = \sum h_{ij}\omega^i\omega^j \tag{5.4.1}$$

as relative metric with respect to the Euclidean normalization. M can be naturally regarded as a Riemannian surface. Choose isothermal parameters u, v with respect to (5.4.1) and let $\xi = u + \sqrt{-1} \cdot v$. The Blaschke metric can be written as

$$G = 2F(\xi, \bar{\xi}) \cdot \mid d\xi \mid^2 \tag{5.4.2}$$

where $d\xi = du + \sqrt{-1} \cdot dv$. As before, denote the components of the cubic Fubini-Pick form (see section 2.4) and the affine Weingarten form (see section 2.3.2), with respect to the local coordinate system (u, v), by A_{ijk} and B_{ij}, respectively. Define α and β by

$$\alpha := \tfrac{1}{2}\left(A_{111} + \sqrt{-1}\, A_{222}\right), \tag{5.4.3}$$

$$\beta := \tfrac{1}{2}\left(\tfrac{B_{11} - B_{22}}{2} - \sqrt{-1}\, B_{12}\right). \tag{5.4.4}$$

Then the Fubini-Pick form and the Weingarten form can be expressed by

$$A = \alpha (d\xi)^3 + \bar{\alpha}(d\bar{\xi})^3, \tag{5.4.5}$$

$$B = \beta (d\xi)^2 + 2FL_1 d\xi d\bar{\xi} + \bar{\beta}(d\bar{\xi})^2. \tag{5.4.6}$$

We use the Cauchy-Riemann operators

$$\tfrac{\partial}{\partial \xi} = \tfrac{1}{2}\left(\tfrac{\partial}{\partial u} - \sqrt{-1} \cdot \tfrac{\partial}{\partial v}\right) \qquad \text{and} \qquad \tfrac{\partial}{\partial \bar{\xi}} = \tfrac{1}{2}\left(\tfrac{\partial}{\partial u} + \sqrt{-1} \cdot \tfrac{\partial}{\partial v}\right).$$

From the theory of complex manifolds, every complex tensor bundle on M is reduced to a direct sum of bigraded complex line bundles $E_{r,s}$, where r and s are integers: locally $E_{r,s}$ is generated by $d\xi^r \otimes d\bar{\xi}^s$; here $d\xi^r$ and $d\bar{\xi}^s$, for r or s negative, denote the contravariant tensors $\left(\frac{\partial}{\partial \xi}\right)^{-r}$ or $\left(\frac{\partial}{\partial \bar{\xi}}\right)^{-s}$, respectively. The tensor products are regarded to be commutative unless specified otherwise. Thus the metric coefficient F is the fibre coordinate of a cross section in $E_{1,1}$. The tangent bundle of M tensored with C splits into the direct sum of $E_{-1,0}$ and $E_{0,-1}$, locally generated by $\frac{\partial}{\partial \xi}$ and $\frac{\partial}{\partial \bar{\xi}}$, respectively.
The Levi-Civita operator ∇ of covariant derivation on smooth sections in $E_{r,s}$ splits into

$$\nabla = \nabla' + \nabla'', \tag{5.4.7}$$

where ∇' is of bidegree (1,0) and ∇'' of bidegree (0,1). They satisfy the following:
(1) ∇, ∇' and ∇'' are linear derivation operators, i.e., for any complex number c and arbitrary smooth sections f, f_1, f_2 in $E_{r,s}$ and g in $E_{p,q}$, each of them, say ∇, satisfies
(i) $\nabla(cg) = c\nabla g$;
(ii) $\nabla(f_1 + f_2) = \nabla f_1 + \nabla f_2$;
(iii) $\nabla(f \otimes g) = (\nabla f) \otimes g + f \otimes (\nabla g)$;

(2) if $f = f(\xi, \bar{\xi})$ and $g = g(\xi, \bar{\xi})$ are the local coefficients of smooth sections in $E_{r,0}$ and in $E_{0,s}$, respectively, then the local coefficients of $\nabla'' f$ in $E_{r,1}$ and of $\nabla' g$ in $E_{1,s}$ are $\frac{\partial f(\xi, \bar{\xi})}{\partial \bar{\xi}}$ and $\frac{\partial g(\xi, \bar{\xi})}{\partial \xi}$;

(3) for scalar f, $\nabla' f = \frac{\partial f}{\partial \xi}$ and $\nabla'' f = \frac{\partial f}{\partial \bar{\xi}}$;

(4) for the metric coefficient F we have $\nabla' F = \nabla'' F = 0$.

From the properties above one can derive the formulas for the general case, where f is a smooth section in $E_{r,s}$:

$$\nabla' f = F^r \frac{\partial}{\partial \xi}(F^{-r} f) = \frac{\partial f}{\partial \xi} - r f \frac{\partial \ln F}{\partial \xi}, \tag{5.4.8}$$

$$\nabla'' f = F^s \frac{\partial}{\partial \bar{\xi}}(F^{-s} f) = \frac{\partial f}{\partial \bar{\xi}} - s f \frac{\partial \ln F}{\partial \bar{\xi}}. \tag{5.4.9}$$

Computing the second covariant derivatives, one obtains the Ricci identity

$$[\nabla', \nabla''] f = \nabla'\nabla'' f - \nabla''\nabla' f = (s - r) F f \kappa, \tag{5.4.10}$$

where as before

$$\kappa = -F^{-1} \frac{\partial^2 \ln F}{\partial \xi \partial \bar{\xi}} \tag{5.4.11}$$

denotes the intrinsic Gaußian curvature of the Blaschke metric. The Laplace operator of the Blaschke metric reads

$$\Delta = \frac{2}{F} \frac{\partial^2}{\partial \xi \partial \bar{\xi}} \,. \tag{5.4.12}$$

Since u and v are isothermal parameters relative to (5.4.1), we have

$$\begin{aligned} & Det\,(x_\xi, x_{\bar{\xi}}, x_{\xi^2}) = Det\,(x_\xi, x_{\bar{\xi}}, x_{\bar{\xi}^2}) = 0, \\ & -\sqrt{-1} \cdot Det\,(x_\xi, x_{\bar{\xi}}, x_{\xi\bar{\xi}}) = F^2 > 0, \end{aligned} \tag{5.4.13}$$

where $x_\xi = \frac{\partial x}{\partial \xi}$, $x_{\bar{\xi}} = \frac{\partial x}{\partial \bar{\xi}}$, $x_{\xi^2} = \frac{\partial^2 x}{\partial \xi \partial \xi}$, etc. The affine normal vector field Y and the conormal vector field U of $x(M)$ satisfy the relations

$$Y = F^{-1} x_{\xi\bar{\xi}}, \tag{5.4.14}$$

$$U = -\sqrt{-1}\, F^{-1}\, [x_\xi, x_{\bar{\xi}}], \tag{5.4.15}$$

where the brackets denote the complex "cross" product (see section 2.1.1) on the complexification $T\mathbb{R}^3 \otimes \mathbb{C}$ of $\mathbb{R}^3$.

$\{x_\xi, x_{\bar{\xi}}, Y\}$ is a complex frame field on $x(M)$, and $\{U, U_\xi, U_{\bar{\xi}}\}$ is a complex frame field on the immersed surface $U : M \to V^*$. These two frame fields satisfy the following relations:

$$Det\,(x_\xi, x_{\bar{\xi}}, Y) = Det\,(U, U_\xi, U_{\bar{\xi}}) = \sqrt{-1}\, F, \tag{5.4.16}$$

$$\begin{pmatrix} U \\ U_\xi \\ U_{\bar{\xi}} \end{pmatrix} \cdot (x_\xi, x_{\bar{\xi}}, Y) = \begin{pmatrix} 0 & 0 & 1 \\ 0 & -F & 0 \\ -F & 0 & 0 \end{pmatrix} \tag{5.4.17}$$

where the operation "$\cdot$" : $(\mathbb{R}^{3*} \otimes \mathbb{C}) \times (\mathbb{R}^3 \otimes \mathbb{C}) \to \mathbb{C}$ denotes the complex inner product. The covariant structure equations in section 2.4.3 can be rewritten in the form

$$\begin{cases} \nabla'\,(x_\xi, x_{\bar{\xi}}, Y) = \left(F^{-1}\alpha\, x_{\bar{\xi}},\ FY,\ -L_1 x_\xi - F^{-1}\beta\, x_{\bar{\xi}}\right), \\[2ex] \nabla''\,(x_\xi, x_{\bar{\xi}}, Y) = \left(FY,\ F^{-1}\bar{\alpha}\, x_\xi,\ -L_1 x_{\bar{\xi}}\ -\ F^{-1}\bar{\beta}\, x_\xi\right), \end{cases} \tag{5.4.18}$$

$$\begin{cases} \nabla'\,(U, U_\xi, U_{\bar{\xi}}) = \left(U_\xi,\ -\beta U - F^{-1}\alpha\, U_{\bar{\xi}},\ -L_1 F U\right), \\[2ex] \nabla''\,(U, U_\xi, U_{\bar{\xi}}) = \left(U_{\bar{\xi}},\ -L_1 F U, -\bar{\beta}\, U - F^{-1}\bar{\alpha} U_\xi\right). \end{cases} \tag{5.4.19}$$

From (5.4.18) and (5.4.19) we obtain

$$\alpha = -\sqrt{-1}\, Det\left(Y, x_\xi, x_{\xi^2}\right) = \sqrt{-1}\, Det\left(U, U_\xi, U_{\xi^2}\right), \tag{5.4.20}$$

$$\beta = \sqrt{-1}\, Det\left(Y, x_\xi, Y_\xi\right) = \sqrt{-1}\, F^{-1} Det\left(U_\xi, U_{\bar\xi}, U_{\xi^2}\right). \tag{5.4.21}$$

In local terms define the forms

$$\hat\alpha := \alpha\,(d\xi)^3 \qquad \text{and} \qquad \hat\beta := \beta\,(d\xi)^2.$$

Obviously, the cubic form $\hat\alpha$ and the quadratic form $\hat\beta$ are independent of the choice of the complex parameters, therefore they are globally defined forms on M. We will call $\hat\alpha$ the Pick form for M. It is easy to see that the zeros of $\hat\alpha$ are the zeros of the Pick invariant, and the zeros of $\hat\beta$ are the umbilic points on M. The Codazzi equations from (2.5.2) and (2.5.4) are expressible as follows

$$\nabla''\alpha = \tfrac{\partial\alpha}{\partial\bar\xi} = -F\beta \qquad \text{and} \qquad \nabla'\bar\alpha = \tfrac{\partial\bar\alpha}{\partial\xi} = -F\bar\beta, \tag{5.4.22}$$

$$\nabla''\beta = \tfrac{\partial\beta}{\partial\bar\xi} = F\tfrac{\partial L_1}{\partial\xi} + F^{-1}\alpha\bar\beta \tag{5.4.23}$$

$$\nabla'\bar\beta = \tfrac{\partial\bar\beta}{\partial\xi} = F\tfrac{\partial L_1}{\partial\bar\xi} + F^{-1}\bar\alpha\beta.$$

The Gauß integrability condition (theorema egregium) reads

$$\kappa = F^{-3}\cdot\alpha\bar\alpha + L_1. \tag{5.4.24}$$

Let $x : M \to \mathbb{R}^3$ be an affine maximal surface and let M be simply connected. Choose a complex isothermal coordinate $\xi = u + \sqrt{-1}\ v$ with respect to the Blaschke metric such that $G = 2F\,|d\xi|^2$. Then the components of its conormal vector field U are harmonic functions, thus there are three holomorphic functions $Z(\xi) = \left(Z(\xi)^1, Z(\xi)^2, Z(\xi)^3\right)$ such that

$$U = \sqrt{-1}\,\left(Z - \bar Z\right). \tag{5.4.25}$$

In the following we shall express quantities of the affine maximal surface in terms of the holomorphic curve $Z(\xi)$ and calculate ΔJ, mainly following Calabi [22]. From (5.4.16), (5.4.19) and (5.4.20) we have

$$F = -\sqrt{-1}\, Det\left(U, U_\xi, U_{\bar\xi}\right) = Det\left(Z - \bar Z, Z', \bar Z'\right) > 0, \tag{5.4.26}$$

$$\alpha = \sqrt{-1}\, Det\left(U, U_\xi, U_{\xi^2}\right) = Det\left(Z - \bar Z, Z', Z''\right), \tag{5.4.27}$$

$$\beta = -F^{-1}\tfrac{\partial\alpha}{\partial\bar\xi} = Det\left(\bar Z', Z', Z''\right)\left[Det\left(Z - \bar Z, Z', \bar Z'\right)\right]^{-1}, \tag{5.4.28}$$

where $Z' = \frac{\partial Z}{\partial\xi}$ and $Z'' = \frac{\partial^2 Z}{\partial\xi^2}$, etc.

Proposition 5.4.1. *Let $x : M \to \mathbb{R}^3$ be a locally strongly convex affine maximal surface and $\xi = u + \sqrt{-1}\,v$ be a local complex isothermal parameter with respect to the Blaschke metric. Then the vector-valued cubic differential form*

$$\Psi = \left(\alpha Y + \beta x_\xi\right)(d\xi)^3$$

and the scalar valued differential form of degree six

$$\Theta = (\beta\nabla'\alpha - \alpha\nabla'\beta)\,(d\xi)^6$$

are holomorphic on M.

To prove the Proposition, we need the following

Lemma 5.4.2. *Let a, b, c, d, e be vectors in $\mathbb{R}^3$ (or in $\mathbb{C}^3$). Then*

$$Det\,(a,b,c)\cdot Det\,(d,b,e) - Det\,(a,b,e)\cdot Det\,(d,b,c) = Det\,(a,b,d)\cdot Det\,(c,b,e)\,.$$

Proof of Proposition 5.4.1. It is sufficient to prove that Ψ and Θ have the following representations in terms of Z

$$\Psi = -\sqrt{-1}\,[Z', Z'']\,(d\xi)^3\,, \qquad\qquad \Theta = Det\,(Z', Z'', Z''')\,(d\xi)^6\,.$$

First, from (5.4.16), (5.4.20) and (5.3.3) we get

$$Y = -\sqrt{-1}\,F^{-1}\,[U_\xi, U_{\bar\xi}]\,,$$

$$x_\xi = \tfrac{1}{2}\left(\tfrac{\partial x}{\partial u} - \sqrt{-1}\,\tfrac{\partial x}{\partial v}\right) = \tfrac{1}{2}\left([U, \tfrac{\partial U}{\partial v}] + \sqrt{-1}\,[U, \tfrac{\partial U}{\partial u}]\right) = \sqrt{-1}\,[U, U_\xi]\,.$$

Hence

$$\alpha Y + \beta x_\xi = -\sqrt{-1}\,\frac{Det(Z-\bar Z, Z', Z'')}{Det(Z-\bar Z, Z', \bar Z')}\,[Z', \bar Z'] - \sqrt{-1}\,\frac{Det(\bar Z', Z', Z'')}{Det(Z-\bar Z, Z', \bar Z')}\,[(Z-\bar Z), Z']\,.$$

Using Lemma 5.4.2, we obtain

$$Det\,(Z-\bar Z, Z', Z'')\,[Z', \bar Z'] - Det\,(Z', \bar Z', \bar Z'')\,[(Z-\bar Z), Z']$$

$$= Det\,(Z-\bar Z, Z', \bar Z')\,[Z', Z'']\,.$$

Consequently we arrive at the following two equations:

$$\Psi = -\sqrt{-1}\,[Z', Z'']\,(d\xi)^3\,,$$

$$\beta\,\nabla'\alpha - \alpha\,\nabla'\beta = -\frac{Det(Z', \bar Z', Z'')\cdot Det(Z-\bar Z, Z', Z''')}{Det(Z-\bar Z, Z', \bar Z')} + \frac{Det(Z-\bar Z, Z', Z'')\cdot Det(-\bar Z', Z', Z''')}{Det(Z-\bar Z, Z', \bar Z')}\,.$$

Again we apply Lemma 5.4.2:

$$-Det\,(Z', \bar Z', Z'')\,Det\,(Z-\bar Z, Z', Z''') + Det\,(Z-\bar Z, Z', Z'')\,Det\,(Z', \bar Z', Z''')$$

$$= Det\,(Z-\bar Z, Z', \bar Z')\,Det\,(Z', Z'', Z''')\,.$$

Hence

$$\Theta = Det\,(Z', Z'', Z''')\,(d\xi)^6\,.$$

∎

Lemma 5.4.3. *The affine Weierstrass representation in section 5.3.1 can be rewritten in terms of the holomorphic curve Z as follows (see [22]):*

$$x = -\sqrt{-1}\left([Z, \bar Z] + \int [Z, dZ] - \int [\bar Z, d\bar Z]\right).$$

Calabi [22] computed the Laplacian of $J = \kappa = F^3 \cdot \alpha\bar{\alpha}$ on affine maximal surfaces; for this computation we again recall the notions of the derivations ∇', ∇'' and basic formulas from (5.4.7) - (5.4.23):

$$\begin{aligned}
\tfrac{1}{2}\Delta J &= F^{-4}\nabla''\nabla'(\alpha\bar{\alpha}) = F^{-4}\,\nabla''\left((\nabla'\alpha)\,\bar{\alpha} + \alpha\left(\overline{\nabla''\alpha}\right)\right) \\
&= F^{-4}\left\{(\nabla'\alpha)\left(\overline{\nabla'\alpha}\right) + (\nabla'\nabla''\alpha + 3F\kappa\,\alpha)\,\bar{\alpha} + \nabla''\left(-\alpha F\bar{\beta}\right)\right\} \\
&= \|\nabla'\alpha\|^2 + 3J^2 + F^{-4}\left[-\nabla''(F\beta)\,\bar{\alpha} - F\left(-F\beta\bar{\beta} + \alpha\left(\overline{\nabla'\beta}\right)\right)\right] \\
&= \|\nabla'\alpha\|^2 + 3J^2 + \|\beta\|^2 - 2F^{-3}Re\,(\bar{\alpha}\,\nabla'\beta)\,,
\end{aligned}$$

where $\|\nabla'\alpha\|^2 = F^{-4}\,(\nabla'\alpha)\left(\overline{\nabla'\alpha}\right)$ and $\|\beta\|^2 = F^{-2}\beta\,\bar{\beta}$. By φ denote the local coefficient of Θ, i.e.,

$$\varphi = \beta\nabla'\alpha - \alpha\nabla'\beta = Det\,(Z', Z'', Z''')\,.$$

When $\alpha \neq 0$, we get the following two relations:

$$\bar{\alpha}\nabla'\beta = \tfrac{\bar{\alpha}}{\alpha}\,(\beta\nabla'\alpha - \varphi)\,,$$

$$\|\nabla'\alpha\|^2 + \|\beta\|^2 - 2F^{-3}Re\left(\frac{\bar{\alpha}}{\alpha}\,(\beta\nabla'\alpha - \varphi)\right) \geq \left\|\nabla'\alpha - \tfrac{F\alpha\bar{\beta}}{\bar{\alpha}}\right\|^2 - 2\,\|\varphi\|\,.$$

Substituting the last inequality into the foregoing calculation of ΔJ, we get

$$\tfrac{1}{2}\Delta J \geq 3J^2 - 2\,\|\varphi\|\,. \tag{5.4.29}$$

(5.4.29) holds at each point where $\alpha \neq 0$. Let p be a point such that $\alpha = 0$. If there is a neighborhood D of p such that $\alpha \equiv 0$ in D, then $\beta \equiv 0$ and so $\varphi \equiv 0$ in D. Hence (5.4.29) holds. If there is a point in every neighborhood of p such that $\alpha \neq 0$, then it follows from a continuity argument that (5.4.29) holds at p. Thus (5.4.29) holds everywhere. ∎

5.4.1 *Computation of* $\Delta\left(J + \|B\|^2\right)$

Applying the above calculation, Calabi computed

$$\Delta\left(J + \|B\|^2\right) = \Delta(\|\alpha\|^2 + 2\,\|\beta\|).$$

Now

$$\begin{aligned}
\Delta\,\|\beta\| &= \left\|\beta^{-1}\right\| \cdot F^{-3}\,\nabla''\nabla'\left(\beta\bar{\beta}\right) - \tfrac{1}{2}\,\|\beta\|^{-3}\,F^{-5}\,\left(\nabla'\left(\beta\bar{\beta}\right)\right)\left(\nabla''\left(\beta\bar{\beta}\right)\right) \\
&= \|\beta\|^{-1}\,F^{-3}\,\left[5F^{-2}\,\alpha\bar{\alpha}\beta\bar{\beta} + 2F^{-1}\,Re\left((\nabla'\alpha)\,\bar{\beta}^2\right) + (\nabla'\beta)\left(\overline{\nabla'\beta}\right)\right] \\
&\quad - \tfrac{1}{2}\,\|\beta\|^{-3}\,F^{-5}\left[\beta\bar{\beta}\,(\nabla'\beta)\left(\overline{\nabla'\beta}\right) + F^{-2}\,\alpha\bar{\alpha}\,\beta^2\bar{\beta}^2 + 2F^{-1}Re\left(\alpha\,(\nabla'\beta)\,\bar{\beta}^3\right)\right] \\
&= \tfrac{9}{2}\,\|\alpha\|^2\,\|\beta\| + \tfrac{1}{2}\,\|\beta\|^{-1}\,\|\beta\|^2 + 2\,\|\beta\|^{-1}\,F^{-4}\,Re\left((\nabla'\alpha)\,\bar{\beta}^2\right) \\
&\quad - F^{-6}\,\|\beta\|^{-3}\,Re\left(\alpha\,(\nabla'\beta)\,\bar{\beta}^3\right).
\end{aligned}$$

It follows from the above calculation of ΔJ that

$$\begin{aligned}\Delta\left(\tfrac{1}{2}J+\|\beta\|\right) &= \Delta\left(\tfrac{1}{2}\|\alpha\|^2+\|\beta\|\right)\\ &= 3\|\alpha\|^4+\left\|\nabla'\alpha+F\beta^{\frac{3}{2}}(\bar\beta)^{-\frac{1}{2}}\right\|^2+\|\beta\|^{-2}\left\|\alpha\bar\beta^{\frac{3}{2}}-\bar\alpha\beta^{\frac{3}{2}}\right\|^2\\ &\quad+\tfrac{1}{2}\|\beta\|^{-1}\left\{\|\nabla'\beta-2\|\beta\|\,\alpha-F^{-1}\bar\alpha\,\beta^2\,\bar\beta^{-1}\|\right\}^2\\ &\geq 3\|\alpha\|^4+\|\beta\|^{-2}\left\|\alpha\bar\beta^{\frac{3}{2}}-\bar\alpha\beta^{\frac{3}{2}}\right\|^2\geq 0.\end{aligned}$$

We use this inequality to prove the following theorem which first was obtained by Martinez and Milan [70]:

Theorem 5.4.4. *Let $x: M\to\mathbb{R}^3$ be a locally strongly convex, affine complete, affine maximal surface. If there is a constant $N>0$ such that the norm of the Weingarten form $\|B\|_G^2$ satisfies $\|B\|_G^2<N$ everywhere on M then $x(M)$ is an elliptic paraboloid.*

Proof. The condition $\|B\|_G^2<N$ implies that $\|\beta\|$ is bounded above by a positive constant C. On the other hand, we have

$$\begin{aligned}\Delta\left(\tfrac{1}{2}J+\|\beta\|\right) &\geq 3J^2\geq 6\cdot\left(\tfrac{1}{2}J+\|\beta\|\right)^2-12\|\beta\|^2\\ &\geq 6\left(\tfrac{1}{2}J+\|\beta\|\right)^2-12C^2.\end{aligned}\tag{5.4.30}$$

We apply Corollary 2.5.10 from ([58], p.125), which implies that $\frac{1}{2}J+\|\beta\|$ is bounded from above. Thus we have a bounded subharmonic function on a complete surface with $\kappa\geq 0$, thus the sum $\frac{1}{2}J+\|\beta\|$ must be a constant. It follows from (5.4.30) that $J=0$ everywhere on M, therefore $x(M)$ is an elliptic paraboloid. ■

5.5 Calabi's Conjecture

In this section we will give a proof of Calabi's conjecture for two dimensions (see [52]). For the proof we use the above Theorem 5.4.4 and a useful Lemma of Hofer [39], which was applied several times in symplectic geometry.

Lemma 5.5.1. ([39], p.535). *Let (X,d) be a complete metric space with metric d, and $\bar B_a(p,d):=\{x\,|\,d(p,x)\leq a\}$ be a ball with center p and radius a. Let Ψ be a non-negative continuous function defined on $\bar B_{2a}(p,d)$. Then there is a point $q\in\bar B_a(p,d)$ and a positive number $\epsilon<\frac{a}{2}$ such that*

$$\Psi(x)\leq 2\Psi(q)\quad\text{for all } x\in\bar B_\epsilon(q,d)\quad\text{and}\quad \epsilon\Psi(q)\geq\tfrac{a}{2}\Psi(p).$$

As in section 5.4, we choose isothermal parameters u, v such that the Blaschke metric is given by $G=F(du^2+dv^2)$, where $F>0$ is a function of u, v. Suppose that M is an affine maximal surface. The formula $\Delta U=-2L_1U$ implies that U is harmonic with respect to u, v. Let $\xi=u+\sqrt{-1}\,v$. Define α and β as in (5.4.3) and (5.4.4), respectively. As before, let $\|\cdot\|_G$ denote the norm with respect to the Blaschke metric, then we have

$$\|\alpha\|^2 := \left(\tfrac{F}{2}\right)^{-3}\cdot\alpha\bar{\alpha} = \tfrac{1}{2}\|A\|_G^2, \quad \|\beta\|^2 := \left(\tfrac{F}{2}\right)^{-2}\cdot\beta\bar{\beta} = \tfrac{1}{2}\|B\|_G^2. \qquad (5.5.1)$$

5.5.1 *Proof of Calabi's Conjecture for dimension* $n = 2$

From the assumption $x : M \to \mathbb{R}^3$ is a locally strongly convex affine maximal surface, which is complete with respect to the Blaschke metric. We want to show that there is a constant $N > 0$ such that $\|B\|_G^2 \le N$ everywhere.

We assume that $\|B\|_G^2$ is not bounded above. Then there is a sequence of points $p_\ell \in M$ such that $\|B\|_G^2(p_\ell) \to \infty$. We may assume that M is simply connected, otherwise we consider its universal covering space. As M is non-compact, but complete with $\kappa = J + L_1 \ge 0$, it is conformally equivalent to the complex plane $\mathbb{C}$. Then we may choose global isothermal parameters u,v on M such that the Blaschke metric is given by $G = F(du^2 + dv^2)$. Let $\bar{B}_1(p_\ell, G)$ be the geodesic ball with center p_ℓ and radius 1. Consider a family $\Psi(\ell) : \bar{B}_2(p_\ell, G) \to \mathbb{R}$ of functions, $\ell \in \mathbb{N}$, defined by

$$\Psi(\ell) = \|\text{grad} \ln F\|_G + \|A\|_G + \|B\|_G^{\frac{1}{2}}.$$

In terms of u, v we have

$$\|\text{grad} \ln F\|_G^2 = \tfrac{1}{F}\left(\left(\tfrac{\partial \ln F}{\partial u}\right)^2 + \left(\tfrac{\partial \ln F}{\partial v}\right)^2\right),$$

$$\|A\|_G^2 = \tfrac{1}{F}\sum (A_{ij}^k)^2, \quad \|B\|_G^2 = \tfrac{1}{F^2}\sum (B_{ij})^2.$$

Using Hofer's Lemma we find a sequence of points q_ℓ and positive numbers ϵ_ℓ such that

$$\Psi(x) \le 2\Psi(q_\ell) \quad \text{for all } x \in \bar{B}_{\epsilon_\ell}(q_\ell, G), \qquad (5.5.2)$$

$$\epsilon_\ell \Psi(q_\ell) \ge \frac{1}{2}\Psi(p_\ell) \to \infty. \qquad (5.5.3)$$

The restriction of the surface x to the balls $\bar{B}_{\epsilon_\ell}(q_\ell, G)$ defines a family $M(\ell)$ of maximal surfaces. For every ℓ, we normalize $M(\ell)$ as follows:

Step 1. Denote by $u(\ell)$, $v(\ell)$ the restriction of the isothermal parameters of M to $M(\ell)$. First we take a parameter transformation:

$$\hat{u}(\ell) = c(\ell)u(\ell), \quad \hat{v}(\ell) = c(\ell)v(\ell), \quad c(\ell) > 0, \qquad (5.5.4)$$

where $c(\ell)$ is a constant. Choosing $c(\ell)$ appropriately and using an obvious notation $\hat{F}$, we may assume that, for every ℓ, we have $\hat{F}(q_\ell) = 1$. Note that, under the parameter transformation (5.5.4), Ψ is invariant.

Step 2. We use the Weierstrass representation for affine maximal surfaces (see section 5.3) to define, for every ℓ, a new surface $\tilde{M}(\ell)$ from $M(\ell)$ via its conormal by

$$\tilde{U}(\ell) = \lambda(\ell)U(\ell), \quad \lambda(\ell) > 0;$$

we introduce new parameters $\tilde{u}(\ell), \tilde{v}(\ell)$ by

$$\tilde{u}(\ell) = b(\ell)\hat{u}(\ell), \quad \tilde{v}(\ell) = b(\ell)\hat{v}(\ell), \quad b(\ell) > 0,$$

where $\lambda(\ell)$ and $b(\ell)$ are constants. From the foregoing conormal equation one easily verifies that each $\tilde{M}(\ell)$ again is a locally strongly convex maximal surface (see section 5.3.1). We now choose $\lambda(\ell) = (b(\ell))^{\frac{2}{3}}$, $b(\ell) = \Psi(q_\ell)$. Using again an obvious notation $\tilde{F}, \tilde{\Psi}$, one can see that

$$\tilde{F} = \hat{F}, \quad \tilde{\Psi}(\ell) = \tfrac{1}{b(\ell)}\Psi(\ell).$$

In fact, the first equation is trivial. Now we calculate the second one. We can easily get

$$\| \operatorname{grad} \ln \tilde{F} \|_{\tilde{G}} = \tfrac{1}{b} \| \operatorname{grad} \ln \hat{F} \|_{\hat{G}} .$$

From (5.5.1), (5.4.20), (5.4.21), (5.4.26) and our choice $\lambda^3 = b^2$ we have

$$\| \tilde{B} \|^2_{\tilde{G}} = 2 \| \tilde{\beta} \|^2 = 2\tfrac{1}{b^4} \| \beta \|^2 = \tfrac{1}{b^4} \| B \|^2_G,$$
$$\| \tilde{A} \|^2_{\tilde{G}} = 2 \| \tilde{\alpha} \|^2 = 2\tfrac{1}{b^2} \| \alpha \|^2 = \tfrac{1}{b^2} \| A \|^2_G .$$

Then the second equality follows.
We denote $\bar{B}_a(q_\ell, \tilde{G}) := \{x \in \tilde{M}(\ell) \,|\, \tilde{r}(\ell)(x, q_\ell) \leq a\}$, where $\tilde{r}(\ell)$ is the geodesic distance function with respect to the Blaschke metric $\tilde{G}$ on $\tilde{M}(\ell)$. Then $\tilde{\Psi}(\ell)$ is defined on the geodesic ball $\bar{B}_{\tilde{\epsilon}(\ell)}(q_\ell, \tilde{G})$ with $\tilde{\epsilon}(\ell) = \epsilon_\ell \Psi(q_\ell) \geq \frac{1}{2}\Psi(p_\ell) \to \infty$. From (5.5.2) we have

$$\tilde{\Psi}(q_\ell) = 1, \qquad \tilde{\Psi}(x) \leq 2, \quad \forall x \in \bar{B}_{\tilde{\epsilon}(\ell)}(q_\ell, \tilde{G}).$$

Step 3. For any ℓ we introduce new parameters $\xi_1(\ell)$, $\xi_2(\ell)$ as follows:

$$\xi_1(\ell) = \tilde{u}(\ell) - \tilde{u}(\ell)(q_\ell), \qquad \xi_2(\ell) = \tilde{v}(\ell) - \tilde{v}(\ell)(q_\ell).$$

Then, at q_ℓ, $(\xi_1, \xi_2) = (0, 0)$ for any ℓ, and we can identify the parametrization (ξ_1, ξ_2) for any index ℓ. Let $\tilde{x}(\ell)$ denote the position vector of $\tilde{M}(\ell)$. An appropriate unimodular affine transformation gives

$$\tilde{x}(\ell)(0) = 0, \tag{5.5.5}$$
$$\tilde{x}_{\xi_1}(\ell)(0) = e_1 = (1, 0, 0), \tag{5.5.6}$$
$$\tilde{x}_{\xi_2}(\ell)(0) = e_2 = (0, 1, 0), \tag{5.5.7}$$
$$\tilde{Y}(\ell)(0) = (0, 0, 1). \tag{5.5.8}$$

Consider the open geodesic balls

$$B_{\tilde{\epsilon}(\ell)}(0, \tilde{G}) := \{(\xi_1, \xi_2) \in \mathbb{R}^2 \,|\, \tilde{r}(\ell)(0, \xi) < \tilde{\epsilon}(\ell)\}$$

and the sequence $\tilde{M}(\ell)$ of maximal surfaces $\tilde{x}(\ell) : B_{\tilde{\epsilon}(\ell)}(0, \tilde{G}) \to \mathbb{R}^3$. They satisfy (5.5.5) - (5.5.8) and the conditions

$$\tilde{F}(\ell)(0) = 1, \tag{5.5.9}$$
$$\tilde{\Psi}(\ell)(0) = 1, \qquad \tilde{\Psi}(\ell)(\xi) \leq 2 \qquad \forall \xi \in B_{\tilde{\epsilon}(\ell)}(0, \tilde{G}), \tag{5.5.10}$$
$$\tilde{\epsilon}(\ell) \to \infty.$$

It follows from (5.3.1) and (5.5.6) - (5.5.9) that, for any ℓ, $(\tilde{U}_{\xi_1}, \tilde{U}_{\xi_2}, \tilde{U})(0) = I$, where I is the unit matrix. We need the following lemma

Lemma 5.5.2. *Let M be an affine maximal surface defined in a neighborhood of $0 \in \mathbb{R}^2$. Suppose that, with the notations from above,*

(i) $$F(0) = 1, \quad (U_{\xi_1}, U_{\xi_2}, U)(0) = I,$$

(ii) $$\left(\tfrac{1}{F}\sum\left(\tfrac{\partial \ln F}{\partial \xi_i}\right)^2\right)^{\frac{1}{2}} + \left(\tfrac{1}{F}\sum (A_{ij}^k)^2\right)^{\frac{1}{2}} + \left(\tfrac{1}{F}\left(\sum (B_{ij})^2\right)^{\frac{1}{2}}\right)^{\frac{1}{2}} \leq 2.$$

Denote $\bar{B}_{\frac{\sqrt{2}}{2}}(0) := \{(\xi_1, \xi_2) \mid \xi_1^2 + \xi_2^2 \leq \frac{1}{2}\}$. Then there is a constant $C_1 > 0$, such that, for $(\xi_1, \xi_2) \in \bar{B}_{\frac{\sqrt{2}}{2}}(0)$, the following estimates hold

(1) $\frac{4}{9} \leq F \leq 4$;
(2) $\|U\| + \|U_{\xi_1}\| + \|U_{\xi_2}\| \leq C_1$, *where $\|\cdot\|$ denotes the canonical norm in $\mathbb{R}^3$;*
(3) *denote $r_o = \frac{1}{3}$; then $\bar{B}_{r_o}(0, G) \subset \{\xi_1^2 + \xi_2^2 < \frac{1}{4}\} \subset \bar{B}_{\frac{\sqrt{2}}{2}}(0)$, where Ω_{r_o} is the geodesic ball with center 0 and radius r_o with respect to the Blaschke metric G.*

Proof. (1) Consider an arbitrary curve
$$\Gamma = \{\xi_1 = a_1 s,\ \xi_2 = a_2 s;\ a_1^2 + a_2^2 = 1,\ s \geq 0\}.$$
By assumption we have
$$\tfrac{1}{F}\left(\tfrac{\partial \ln F}{\partial s}\right)^2 \leq 2, \qquad F(0) = 1.$$
Solving this differential inequality with $F(0) = 1$, we get
$$\left(\frac{1}{1+\frac{\sqrt{2}}{2}s}\right)^2 \leq F(s) \leq \left(\frac{1}{1-\frac{\sqrt{2}}{2}s}\right)^2.$$
From the assumption we have $s \leq \frac{1}{\sqrt{2}}$, then **(1)** follows.
(2) Note that the Christoffel symbols are given by $\frac{\partial \ln F}{\partial \xi_i}$. Along the curve Γ the structure equation $U_{,ij} = -\sum A_{ij}^k U_{,k} - B_{ij} U$ gives an ODE, which can be written in matrix form:
$$\frac{dX}{ds} = XD, \tag{5.5.11}$$
where $X = (U_{\xi_1}, U_{\xi_2}, U)$, and D is a matrix, whose elements depend on B_{ij}, A_{ij}^k and $\frac{\partial \ln F}{\partial \xi_i}$. From (5.5.11) it follows that
$$\frac{dX^t}{ds} = D^t X^t, \tag{5.5.12}$$
where we use an obvious notation for the transpose of a matrix. Then
$$\frac{d(X^t X)}{ds} = D^t X^t X + X^t X D. \tag{5.5.13}$$
Denote $f := tr(X^t X)$. Taking the trace of (5.5.13) we get
$$\frac{df}{ds} = tr(D^t X^t X) + tr(X^t X D) \leq Cf, \tag{5.5.14}$$

where C is a constant. Deriving the last inequality we use **(1)** and the condition (ii). Solving (5.5.14) with the condition (i) we get **(2)**.
(3) From **(1)** we immediately get **(3)**. ∎

We continue with the proof of Calabi's conjecture. Since $\tilde{\epsilon}(\ell) \to \infty$, we have the relation $\bar{B}_{\frac{\sqrt{2}}{2}}(0) \subset B_{\tilde{\epsilon}(\ell)}(0, \tilde{G})$ for ℓ big enough. In fact, by **(1)**, the geodesic distance from 0 to the boundary of $\bar{B}_{\frac{\sqrt{2}}{2}}(0)$ with respect to the Blaschke metric on $\tilde{M}(\ell)$ is less than $\sqrt{2}$. Using **(2)** and a standard elliptic estimate, we get a C^k-estimate, independent of ℓ, for any k. It follows that there is a ball $\{\xi_1^2 + \xi_2^2 \leq C_2\}$ and a subsequence (still indexed by ℓ) such that $\tilde{U}(\ell)$ converges to $\tilde{U}$ on the ball, and correspondingly all derivatives, where $C_2 < \frac{1}{2}$ is close to $\frac{1}{2}$. Thus, as limit, we get a maximal surface $\tilde{M}$, defined on the ball, which contains a geodesic ball $\bar{B}_{r_o}(0, \tilde{G})$. We now extend the surface $\tilde{M}$ as follows: For every boundary point $p = (\xi_{1o}, \xi_{2o})$ of the geodesic ball $\bar{B}_{r_o}(0, \tilde{G})$ we first make the parameter transformation:
$\tilde{\xi}_i = b(\xi_i - \xi_{io})$ such that, at p, $(\tilde{\xi}_1, \tilde{\xi}_2) = (0, 0)$, and for the limit surface $\tilde{M}$ we have $\tilde{F}(p) = 1$. We choose a frame e_1, e_2, e_3 at p such that $e_1 = \tilde{x}_{\tilde{\xi}_1}$, $e_2 = \tilde{x}_{\tilde{\xi}_2}$, $e_3 = \tilde{Y}$. We have

$$\tilde{F}(\ell)(p) \to \tilde{F}(p) = 1, \quad \left(\tilde{U}_{\tilde{\xi}_1}(\ell), \tilde{U}_{\tilde{\xi}_2}(\ell), \tilde{U}(\ell)\right)(p) \to I \quad \text{as} \quad \ell \to \infty. \qquad (i')$$

It is easy to see that, under the conditions (i') and (ii) in Lemma 5.5.2, the estimates **(1)**, **(2)** and **(3)** in Lemma 5.5.2 hold again. By the same argument as above we conclude that there is a ball around p and a subsequence ℓ_k, such that $\tilde{U}(\ell)$ converges to $\tilde{U}'$ on the ball, and correspondingly all derivatives. As limit, we get a maximal surface $\tilde{M}'$, which contains a geodesic ball of radius r_o around p. Then we return to the original parameters ξ_1, ξ_2 and the original frame e_1, e_2, e_3 at 0. Note that the geodesic distance is independent of the choice of the parameters and the frames. It is obvious that $\tilde{M}$ and $\tilde{M}'$ agree on the common part. We repeat this procedure to extend $\tilde{M}$ to be defined on $\bar{B}_{2r_o}(0, \tilde{G})$, etc. In this way we may extend $\tilde{M}$ to be an affine complete maximal surface defined in a domain $\Omega \in \mathbb{R}^2$; using (5.5.9) and (5.5.10) we get

$$\| \tilde{B} \|_{\tilde{G}} \leq 4, \quad \tilde{\Psi}(0) = 1.$$

By Theorem 5.4.4, $\tilde{M}$ must be an elliptic paraboloid, given by

$$x^3 = \frac{1}{2}((x^1)^2 + (x^2)^2), \qquad (5.5.15)$$

where x^1, x^2, x^3 are the coordinates in $\mathbb{R}^3$ with respect to the frame e_1, e_2, e_3. For a paraboloid we have $\| \tilde{A} \|_{\tilde{G}} = 0$, $\| \tilde{B} \|_{\tilde{G}}^2 = 0$, $\tilde{R} = 0$ identically, and $\tilde{G} = (dx^1)^2 + (dx^2)^2$. Thus

$$\| \operatorname{grad} \ln \tilde{F} \|_{\tilde{G}} (0) = 1. \qquad (5.5.16)$$

We consider $\ln \tilde{F}$ as a function of x^1, x^2. Since the scalar curvature vanishes identically, $\tilde{R} = 0$, from the formula

$$\Delta \ln \tilde{F} = -\tilde{R}$$

we conclude that $\ln \tilde{F}$ is a harmonic function. As $\| \operatorname{grad} \ln \tilde{F} \| \leq 2$, $\ln \tilde{F}$ must be a linear function. In view of (5.5.15), without loss of generality, we may assume that $\ln \tilde{F} = x^1$. We introduce complex coordinates and write $w = \xi_1 + \sqrt{-1}\,\xi_2$, $z = x^1 + \sqrt{-1}\,x^2$, then $w(z)$ is a holomorphic or anti-holomorphic function. We consider the case that w is holomorphic. For the case that w is anti-holomorphic, the discussion is similar. Since $\tilde{G} = |dz|^2 = \tilde{F}|dw|^2$, we have $|w'|^2 = \tilde{F}^{-1} = e^{-x^1}$. Let $Q = e^{\frac{z}{2}}$. Then $|w'Q| = 1$. From the maximum principle we get $w'Q = C$, where C is a constant with $|C| = 1$. So $w' = Ce^{-\frac{z}{2}}$. It follows that $w = -2Ce^{-\frac{z}{2}} + E$, where E is a constant. Since $e^{-\frac{z}{2}}$ has period 2π for x^2, we have a covering map $\mathbb{R}^2 \to \Omega$; this is impossible. We get a contradiction. So $\|B\|_G$ must be bounded above on M. By Theorem 5.4.4, M is an elliptic paraboloid. We have proved Calabi's Conjecture in dimension $n = 2$. ■

5.6 Chern's Conjecture

In this section we study a nonlinear, fourth order partial differential equation for a convex function f on a convex domain Ω in $\mathbb{R}^n$. The equation can be written as

$$\sum_{i,j=1}^{n} F^{ij} w_{ij} = -L^{\sharp}, \qquad w := \left[\det\left(\frac{\partial^2 f}{\partial x^i \partial x^j}\right)\right]^a, \tag{5.6.1}$$

where $L^{\sharp} : \Omega \to \mathbb{R}$ is some given C^∞ function, (F^{ij}) denotes the cofactor matrix of the Hessian matrix $\left(\frac{\partial^2 f}{\partial x^i \partial x^j}\right)$ and $a \neq 0$ is a real constant.

In the case when $a = -\frac{n+1}{n+2}$ and $L^{\sharp} = 0$, the PDE (5.6.1) is the equation for affine maximal hypersurfaces (see section 5.2.2).

In the case when $a = -1$, the PDE (5.6.1) is called the *Abreu equation*, which appears in the study of the differential geometry of toric varieties (see [1], [29], [30], [31]), where $L^{\sharp}$ is the scalar curvature of the Kähler metric. About the Bernstein property for the Abreu equation we would like to pose the following conjecture:

Conjecture 5.6.1. *Let $f(x^1, ..., x^n)$ be a smooth, strictly convex function defined for all $x \in \mathbb{R}^n$. Assume that f satisfies the Abreu equation*

$$\sum_{i,j=1}^{n} F^{ij} w_{ij} = 0, \qquad w := \left[\det\left(\frac{\partial^2 f}{\partial x^i \partial x^j}\right)\right]^{-1}.$$

Then f must be a quadratic polynomial.

Note that the PDE (5.6.1) with $L^{\sharp} = 0$ is the Euler-Lagrange equation of a volume variational problem. In fact, let $f(x)$ be a smooth, strictly convex function defined in a convex domain $\Omega \in \mathbb{R}^n$, then

$$M := \{(x^1, \cdots, x^n, f(x)) \mid (x^1, ..., x^n) \in \Omega\}$$

is a locally strongly hypersurface immersed in $\mathbb{R}^{n+1}$. As in section 3.3.5, for the above graph, we introduce the α-metric

$$G^{(\alpha)} = \rho^{\alpha}\mathfrak{H},$$

where $\mathfrak{H}$ and ρ are defined in section 3.3.4, and $\alpha = -\frac{(n+2)(2a+1)}{n}$; then the $G^{(\alpha)}$-volume is given by

$$V(f,\alpha) = \int_{\Omega}\left[\det\left(\frac{\partial^2 f}{\partial x^i \partial x^j}\right)\right]^{a+1} dx^1 \wedge \cdots \wedge dx^n. \tag{5.6.2}$$

Let $f_t(x) = f(x) + t\varphi(x)$, where $\varphi(x) \in C_0^{\infty}(\Omega)$, then

$$\begin{aligned}\frac{d}{dt}|_{t=0}V(f_t,\alpha) =&(a+1)\int_{\Omega}[\det(f_{ij})]^a \sum F^{ij}\varphi_{ij}\, dx^1\cdots dx^n\\ =&-a(a+1)\int_{\Omega}[\det(f_{ij})]^{a-1}\sum F^{ij}\varphi_i[\det(f_{ij})]_j\, dx^1\cdots dx^n\\ =&(a+1)\int_{\Omega}\left(\sum F^{ij}w_{ij}\right)\varphi\, dx^1\cdots dx^n.\end{aligned}$$

It is easy to see that, in case that $a \neq 0$ and $a \neq -1$, if f is a critical point under any interior variation then

$$\sum_{i,j=1}^{n} F^{ij}w_{ij} = 0,$$

where w was defined in (5.6.1).
Denote by Δ and $\|\cdot\|$ the Laplacian and the tensor norm with respect to the Calabi metric $\mathfrak{H}$, respectively. Recall the definition of ρ from section 1.4. In terms of the Calabi metric the PDE (5.6.1) can be rewritten as

$$\Delta\rho = -\beta\frac{\|\text{grad }\rho\|^2}{\rho} + \frac{1}{a(n+2)}L^{\sharp}\rho^{(a+1)(n+2)+1}, \tag{5.6.3}$$

where

$$\beta := -\frac{(n+2)(2a+1)+2}{2}.$$

Note that the PDE (5.6.3) with $L^{\sharp} = 0$ includes the cases $a = 0$, $a = -1$.

In this section we shall prove

Theorem 5.6.2. [55]. *Let f be a smooth, strictly convex function defined for all $(x_1, x_2) \in \mathbb{R}^2$. If f satisfies the PDE*

$$\sum_{i,j=1}^{2} F^{ij}w_{ij} = 0 \tag{5.6.4}$$

with w from (5.6.1) and $a \leq -\frac{3}{4}$ then f must be a quadratic polynomial.

Remark.
(1) When $a = -\frac{3}{4}$ Theorem 5.6.2 gives a new analytic proof for Chern's conjecture on affine maximal surfaces.

(2) When $a = -1$ Theorem 5.6.2 solves the above Conjecture 5.6.1 affirmatively for $n = 2$.

(3) In [93] Trudinger and Wang proved that the global solution of the PDE (5.6.4) on $\mathbb{R}^2$ with $a > 0$ must be a quadratic polynomial.

(4) When $a = 0$, $n = 2$ and $L^\sharp = 0$ the PDE (5.6.3) reduces to $\Delta\rho = 3\frac{\|\text{grad }\rho\|^2}{\rho}$, which is equivalent to

$$\sum_{i,j=1}^{2} f^{ij}\frac{\partial^2}{\partial x^i \partial x^j}[\ln\det(f_{kl})] = 0. \tag{5.6.5}$$

The global solution of the PDE (5.6.5) on $\mathbb{R}^2$ is not unique. In fact, the two examples in section 4.5.1, restricted to $n = 2$, are global solutions of (5.6.5).

To prove Theorem 5.6.2, we will derive a series of estimates in the subsections 5.6.1 - 5.6.4. The proof of Theorem 5.6.2 follows in subsection 5.6.5.

5.6.1 *Technical estimates*

Let Ω be a convex domain and $0 \in \Omega$ be the center of Ω (for the definition of the center of a bounded convex domain please see section 1.8 in [37]). Let f be a strictly convex function defined on $\Omega \subset \mathbb{R}^n$. Assume that

$$\inf_\Omega f = 0, \qquad f = C > 0 \qquad \text{on } \partial\Omega.$$

In the following we use the Calabi metric $\mathfrak{H}$. Consider the function

$$F := \exp\left\{\tfrac{-m}{C-f} + \tau^\sharp\right\} Q\|\text{grad } h\|^2, \tag{5.6.6}$$

where $Q > 0$, $\tau^\sharp > 0$ and h are smooth functions defined on $\overline{\Omega}$. Clearly, F attains its supremum at some interior point p^*. We choose a local orthonormal frame field such that, at p^*, $h_{,1} = \|\text{grad } h\|$, $h_{,i} = 0$, for all $i > 1$. By the same calculation as in the proof of Lemma 4.5.6, we get the following Lemma (for details see [55]):

Lemma 5.6.3. *At the point p^*, we have the following estimate:*

$$\begin{aligned}
2\left(\tfrac{1}{n-1} - \delta - 1\right)&(h_{,11})^2 + 2\sum h_{,j}(\Delta h)_{,j} + 2(1-\delta)\sum A^2_{ml1}(h_{,1})^2 \\
&- \tfrac{(n+2)^2}{8\delta}\Phi(h_{,1})^2 - \tfrac{2}{\delta(n-1)^2}(\Delta h)^2 \\
&+ \left[\Delta\tau^\sharp - g'\sum(f_{,i})^2 - g\Delta f + \tfrac{\Delta Q}{Q} - \tfrac{\sum(Q_{,i})^2}{Q^2}\right](h_{,1})^2 \le 0
\end{aligned} \tag{5.6.7}$$

for any $0 < \delta < 1$, where here and later

$$g := \tfrac{m}{(C-f)^2}, \qquad g' := 2\tfrac{m}{(C-f)^3}.$$

Consequences. In the following we calculate the expression

$$2\sum h_{,j}(\Delta h)_{,j} + 2(1-\delta)\sum A_{mli}A_{mlj}h_{,i}h_{,j}$$

for the cases $h = f$ and $h = \xi_1$, respectively.

1. The case $h = f$.

Using the formula (4.5.5) we have

$$2\sum f_{,j}(\Delta f)_{,j} = (n+2)\left[\tfrac{\rho_{,11}}{\rho}(f_{,1})^2 - \tfrac{(\rho_{,1})^2}{\rho^2}(f_{,1})^2 + \sum f_{,1i}f_{,1}\tfrac{\rho_{,i}}{\rho}\right]. \qquad (5.6.8)$$

Note that

$$f_{,ij} = A_{ij1}f_{,1} + f_{ij}$$

and

$$\sum (f_{,ij})^2 = \sum A_{ij1}^2(f_{,1})^2 + n + (n+2)\tfrac{\rho_{,1}f_{,1}}{\rho}. \qquad (5.6.9)$$

Similar to (4.5.43) we get

$$\sum (f_{,ij})^2 \geq (\tfrac{n}{n-1} - \delta)(f_{,11})^2 + 2\sum_{i>1}(f_{,1i})^2 - \tfrac{1}{\delta(n-1)^2}(\Delta f)^2$$

for any $0 < \delta < 1$. Thus combination of (5.6.8) and (5.6.9) gives (recall the definition of Φ from section 1.4)

$$\begin{aligned} 2\sum f_{,j}(\Delta f)_{,j} &+ 2(1-\delta)\sum A_{ml1}^2(f_{,1})^2 \\ &\geq \tfrac{2n-5n\delta}{n-1}(f_{,11})^2 + (4-8\delta)\sum_{i>1}(f_{,1i})^2 - \tfrac{2}{\delta(n-1)^2}(\Delta f)^2 \\ &+ (n+2)\tfrac{\rho_{,11}}{\rho}(f_{,1})^2 - \left(\tfrac{(n+2)^2}{4\delta} + (n+2)\right)\Phi(f_{,1})^2 \\ &- 2n - 2(n+2)(1-\delta)\tfrac{\rho_{,1}f_{,1}}{\rho} \end{aligned} \qquad (5.6.10)$$

for any $0 < \delta < 1$. In particular, for $n = 2$, we have

$$2\sum f_{,j}(\Delta f)_{,j} + 2(1-\delta)\sum A_{ml1}^2(f_{,1})^2$$

$$\geq 4\tfrac{\rho_{,11}}{\rho}(f_{,1})^2 + (4-10\delta)\sum (f_{,1i})^2 - \left(\tfrac{4}{\delta}+8\right)\Phi(f_{,1})^2 - \tfrac{2}{\delta}(\Delta f)^2 - 8. \qquad (5.6.11)$$

2. The case $h = \xi_1$.

By (3.3.8) we have

$$\Delta h = -\tfrac{n+2}{2}\mathfrak{H}(\text{grad}\ln\rho,\ \text{grad}\, h).$$

It follows that

$$\begin{aligned} 2\sum h_{,j}(\Delta h)_{,j} &= -(n+2)\left[\tfrac{\rho_{,11}}{\rho}(h_{,1})^2 - \tfrac{(\rho_{,1})^2}{\rho^2}(h_{,1})^2 + \sum h_{,1i}h_{,1}\tfrac{\rho_{,i}}{\rho}\right] \\ &\geq -(n+2)\tfrac{\rho_{,11}}{\rho}(h_{,1})^2 - \delta\sum (h_{,1i})^2 - \tfrac{(n+2)^2}{4\delta}\Phi(h_{,1})^2 \end{aligned}$$

for any $0<\delta<\frac{1}{3}$. Now we use the coordinates $x^1,...,x^n$ to calculate $\sum(h_{,ij})^2$ and $\sum A^2_{ml1}(h_{,1})^2$. Recall that the Christoffel symbols satisfy $\Gamma^k_{ij}=\frac{1}{2}\sum f^{kl}f_{lij}$. Hence

$$h_{,ij}=f_{1ij}-\tfrac{1}{2}\sum f_{1k}f^{kl}f_{lij}=\tfrac{1}{2}f_{1ij},$$
$$\sum(h_{,ij})^2=\tfrac{1}{4}\sum f^{ik}f^{jl}f_{1ij}f_{1kl},$$
$$\sum A^2_{ml1}(h_{,1})^2=\tfrac{1}{4}\sum f^{ik}f^{jl}f_{ijp}f_{klq}f^{pr}f_{1r}f^{qs}f_{1s}=\sum(h_{,ij})^2.$$

Consequently, we obtain

$$\begin{aligned}2\sum h_{,j}(\Delta h)_{,j}&+2(1-\delta)\sum A_{mli}A_{mlj}h_{,i}h_{,j}\\ &\geq(2-3\delta)\sum(h_{,ij})^2-(n+2)\tfrac{\rho_{,11}}{\rho}(h_{,1})^2-\tfrac{(n+2)^2}{4\delta}\Phi(h_{,1})^2\\ &\geq\tfrac{2n-5n\delta}{n-1}(h_{,11})^2+(4-6\delta)\sum_{i>1}(h_{,1i})^2-(n+2)\tfrac{\rho_{,11}}{\rho}(h_{,1})^2\\ &\quad-\left(\tfrac{(n+2)^2}{4\delta}+\tfrac{(n+2)^2}{\delta(n-1)^2}\right)\Phi(h_{,1})^2 \qquad (5.6.12)\end{aligned}$$

for any $0<\delta<\frac{1}{3}$.

5.6.2 *Estimates for the determinant of the Hessian*

In this subsection we shall estimate the determinant of the Hessian of certain functions from above. For this we use $\|\cdot\|_E$ to denote the norm of a vector with respect to the canonical Euclidean metric in $\mathbb{R}^n$. For affine maximal hypersurfaces, Trudinger and Wang [91] obtained upper bounds for the determinant of the Hessian.

Lemma 5.6.4. *Let f be a nonnegative convex C^∞ function defined on the section $S_f(C)$, satisfying the PDE (4.5.9). If $\beta>-\frac{n+4}{2}$ then the following estimate holds:*

$$\det(f_{ij})\leq b_0 \qquad for \quad x\in S_f(C'),$$

where b_0 is a constant depending only on β, $C>0$, $\frac{C'}{C}<1$ and $max\|grad\, f\|_E$.

Proof. Consider the function

$$F:=\exp\left\{\tfrac{-m}{C-f}+\epsilon\sum(\xi_k)^2\right\}\frac{1}{\rho}$$

defined on the *section* $S_f(C)$, where m and ϵ are positive constants to be determined later. Clearly, F attains its supremum at some interior point p^* of $S_f(C)$. We choose a local orthonormal frame field of the metric $\mathfrak{H}$ on M near p^*. Then, at p^*,

$$-gf_{,i}+\epsilon\left(\sum(\xi_k)^2\right)_{,i}-\tfrac{\rho_{,i}}{\rho}=0, \qquad (5.6.14)$$

$$\begin{aligned}-g'\sum(f_{,i})^2-ng-\tfrac{n+2}{2}\sum\left(gf_{,i}+\epsilon\left(\sum(\xi_k)^2\right)_{,i}\right)\tfrac{\rho_{,i}}{\rho}&\\ +2\epsilon\sum u^{ii}+(1+\beta)\Phi\leq0,&\qquad (5.6.15)\end{aligned}$$

where g, g' were defined in Lemma 5.6.3. We inserting (5.6.14) into (5.6.15):

$$2\epsilon\sum u^{ii} - g'\sum (f_{,i})^2 - ng + (\tfrac{n+4}{2}+\beta)\Phi - (n+2)\epsilon\sum \tfrac{\rho_{,i}}{\rho}\left(\sum(\xi_k)^2\right)_{,i} \le 0. \quad (5.6.16)$$

Using the inequality of Schwarz yields

$$(n+2)\epsilon\sum \tfrac{\rho_{,i}}{\rho}\left(\sum(\xi_k)^2\right)_{,i} \le \tfrac{8(n+2)^2\epsilon^2}{n+4+2\beta}\|\text{grad } f\|_E^2\sum u^{ii} + \tfrac{n+4+2\beta}{8}\Phi.$$

From (5.6.14) we have

$$\Phi \ge \tfrac{1}{2}g^2\sum(f_{,i})^2 - 4\epsilon^2\|\text{grad } f\|_E^2\sum u^{ii}.$$

Then

$$\left(2\epsilon - \left[(n+4+2\beta) + \tfrac{8(n+2)^2}{n+4+2\beta}\right]\epsilon^2\|\text{grad } f\|_E^2\right)\sum u^{ii}$$
$$+(\tfrac{n+4+2\beta}{8}g^2 - g')\sum(f_{,i})^2 - ng \le 0.$$

Choose ϵ such that

$$\left[(n+4+2\beta) + \tfrac{8(n+2)^2}{n+4+2\beta}\right]\max\|\text{grad } f\|_E^2 \cdot \epsilon \le 1$$

and $m = \frac{16C}{n+4+2\beta}$. Note that $\sum u^{ii} \ge n\rho^{-\frac{n+2}{n}}$. We use the inequality of Schwarz and get

$$\tfrac{1}{\rho} \le a_0 g^{\frac{n}{n+2}}$$

for some constant a_0 depending only on β, n and $\max\|\text{grad } f\|_E$. Thus we complete the proof of Lemma 5.6.4. ■

Lemma 5.6.5. *Let u be a nonnegative convex C^∞ function defined on the section $S_u(C)$, satisfying the PDE (4.5.9). If $\beta < \frac{n}{2}$ then the following estimate holds:*

$$\det(u_{ij}) \le b_0, \qquad \textit{for } \ \xi \in S_u(C'),$$

where b_0 is a constant depending only on β, $C > 0$, $\frac{C'}{C} < 1$ and $\max\|\text{grad } u\|_E$.

Proof. Consider the function

$$F := \exp\left\{\tfrac{-m}{C-u} + \epsilon\sum(x^k)^2\right\}\rho$$

defined on the *section* $S_u(C)$, where m and ϵ are positive constants to be determined later. Clearly, F attains its supremum at some interior point p^* of $S_u(C)$. We choose a local orthonormal frame field of the metric $\mathfrak{H}$ on M near p^*. Then, at p^*,

$$-\gamma u_{,i} + \epsilon\left(\sum(x^k)^2\right)_{,i} + \tfrac{\rho_{,i}}{\rho} = 0, \quad (5.6.17)$$

$$-\gamma'\sum(u_{,i})^2 - n\gamma + \tfrac{n+2}{2}\sum\tfrac{\rho_{,i}}{\rho}\left(\gamma u_{,i} + \epsilon\left(\sum(x^k)^2\right)_{,i}\right)$$
$$+ 2\epsilon\sum f^{ii} - (1+\beta)\Phi \le 0, \quad (5.6.18)$$

where γ and γ' are the functions defined in section 4.5.4. Inserting (5.6.17) into (5.6.18) yields

$$(n+2)\epsilon \sum \tfrac{\rho_{,i}}{\rho} \left(\sum (x^k)^2\right)_{,i} - \gamma' \sum (u_{,i})^2 - n\gamma + 2\epsilon \sum f^{ii} + \left(\tfrac{n}{2} - \beta\right) \Phi \le 0.$$

Again we use the inequality of Schwarz and get

$$(n+2)\epsilon \sum \tfrac{\rho_{,i}}{\rho} \left(\sum (x^k)^2\right)_{,i} \le \tfrac{8(n+2)^2 \epsilon^2}{n-2\beta} \|\text{grad } u\|_E^2 \sum f^{ii} + \tfrac{n-2\beta}{8} \Phi.$$

From (5.6.17) we have

$$\Phi \ge \tfrac{1}{2}\gamma^2 \sum (u_{,i})^2 - 4\epsilon^2 \|\text{grad } u\|_E^2 \sum f^{ii}.$$

Then

$$\left(2\epsilon - \left[(n-2\beta) + \tfrac{8(n+2)^2}{n-2\beta}\right] \epsilon^2 \|\text{grad } u\|_E^2\right) \sum f^{ii} + \left(\tfrac{n-2\beta}{8}\gamma^2 - \gamma'\right) \sum (u_{,i})^2 - n\gamma \le 0.$$

Choose

$$\left[(n-2\beta) + \tfrac{8(n+2)^2}{n-2\beta}\right] \max \|\text{grad } u\|_E^2 \cdot \epsilon \le 1$$

and $m = \frac{16C}{n-2\beta}$. The inequality of Schwarz gives

$$\rho \le a_0 \gamma^{\frac{n}{n+2}}$$

for some constant a_0 depending only on β, n and $\max \|\text{grad } u\|_E$. Thus we complete the proof of Lemma 5.6.5. ∎

In the following we estimate the determinant $\det\left(\frac{\partial^2 f}{\partial x^i \partial x^j}\right)$ from below and above. We restrict to functions f of two variables. In this case we can estimate the determinant in a convex domain $\Omega \subset \mathbb{R}^2$, while usual estimates hold only in *sections*, just like in Lemmas 5.6.4 and 5.6.5. For simplicity, we restrict to the case $L^\sharp = 0$ in (5.6.3). In fact, the following Lemmas 5.6.6 and 5.6.7 hold for $L^\sharp = const$, see Lemmas 6.1.4 and 6.1.5 below, or [53].

Lemma 5.6.6. *Let f be a smooth, strictly convex function defined on a bounded convex domain $\Omega \subset \mathbb{R}^2$, satisfying the PDE (4.5.9). As before denote by Ω^* the Legendre transformation domain of f. Let Ω'^* be an arbitrary subdomain of Ω^* with $dist(\Omega'^*, \partial\Omega^*) > 0$. Then the following estimate holds:*

$$\det(f_{ij}) \ge b_0 \quad for \quad \xi \in \Omega'^*,$$

where b_0 is a constant depending only on $dist(\Omega'^*, \partial\Omega^*)$, $diam(\Omega)$, $diam(\Omega^*)$ and β.

Proof. For any $\xi_0 = (\xi_1^0, \xi_2^0) \in \Omega^*$ choose $\delta > 0$ such that $0 < \delta < \text{dist}(\xi_0, \partial\Omega^*)$. By an orthogonal transformation we may assume that $\xi_0 = 0$. Consider the function

$$F := -\frac{m}{(\delta^2 - \theta)^\ell} + \rho^k \left(1 + \epsilon J^\sharp\right)$$

defined on $B_\delta^*(0)$, where $\theta := \sum (\xi_k)^2$, $J^\sharp := \sum (x^k)^2$ and m, k, ℓ and ϵ are positive constants to be determined later. Clearly, F attains its supremum at some interior point ξ^* of $B_r^*(0)$. We choose a local orthonormal frame field e_1, e_2 of the metric $\mathfrak{H}$ on M near ξ^*. Then, at ξ^*,

$$-\tfrac{m\ell}{(\delta^2-\theta)^{\ell+1}}\theta_{,i} + \epsilon\rho^k J^\sharp_{,i} + k\rho^{k-1}(1+\epsilon J^\sharp)\rho_{,i} = 0, \tag{5.6.19}$$

$$-\tfrac{m\ell(\ell+1)}{(\delta^2-\theta)^{\ell+2}}\sum(\theta_{,i})^2 - \tfrac{m\ell}{(\delta^2-\theta)^{\ell+1}}\Delta\theta + \epsilon\rho^k\Delta J^\sharp + k\rho^{k-1}(1+\epsilon J^\sharp)\Delta\rho$$

$$+k(k-1)\rho^{k-2}(1+\epsilon J^\sharp)\sum(\rho_{,i})^2 + 2\epsilon k\rho^{k-1}\sum\rho_{,i}J^\sharp_{,i} \le 0. \tag{5.6.20}$$

Using the formulas (3.3.6) and (3.3.7), we obtain

$$-\tfrac{4m\ell(\ell+1)}{(\delta^2-\theta)^{\ell+2}}\sum u^{ij}\xi_i\xi_j - \tfrac{2m\ell}{(\delta^2-\theta)^{\ell+1}}\sum u^{ii} + \tfrac{2m\ell}{(\delta^2-\theta)^{\ell+1}}\sum\tfrac{\rho_{,i}}{\rho}\theta_{,i} + 2\epsilon\rho^k\sum f^{ii}$$
$$+\left(k(k-1)-k\beta\right)\rho^{k-2}(1+\epsilon J^\sharp)\sum(\rho_{,i})^2 + 2\epsilon(k+1)\rho^{k-1}\sum\rho_{,i}J^\sharp_{,i} \le 0.$$

By (5.6.19)

$$\begin{aligned}\tfrac{2m\ell}{(\delta^2-\theta)^{\ell+1}}\sum\tfrac{\rho_{,i}}{\rho}\theta_{,i} &\ge -\tfrac{2\epsilon}{k}\tfrac{m\ell}{(\delta^2-\theta)^{\ell+1}}\sum J^\sharp_{,i}\theta_{,i}\\ &\ge -\tfrac{8\epsilon}{k}\tfrac{m\ell}{(\delta^2-\theta)^{\ell+1}}\,\delta\,\mathrm{diam}(\Omega).\end{aligned}$$

Here we used the fact $\sum J^\sharp_{,i}\theta_{,i} = 4\sum x^i\xi_i \le 4\delta\cdot\mathrm{diam}(\Omega)$. We choose a positive number k, depending only on β, such that $k(k-1-\beta)\ge 1$. The inequality of Schwarz gives

$$-\tfrac{4m\ell(\ell+1)}{(\delta^2-\theta)^{\ell+2}}\sum u^{ij}\xi_i\xi_j - \tfrac{2m\ell}{(\delta^2-\theta)^{\ell+1}}\sum u^{ii} + 2\epsilon\rho^k\sum f^{ii}$$
$$-4\epsilon^2(k+1)^2\rho^k\sum f^{ij}x^ix^j - \tfrac{8\epsilon}{k}\tfrac{m\ell}{(\delta^2-\theta)^{\ell+1}}d(\Omega^*)\,d(\Omega) \le 0, \tag{5.6.21}$$

where $d(\Omega^*)$ and $d(\Omega)$ denote $\mathrm{diam}(\Omega^*)$ and $\mathrm{diam}(\Omega)$, respectively. Choose ϵ such that $4\epsilon(k+1)^2 d(\Omega)^2 \le 1$. By λ_1, λ_2 denote the eigenvalues of $\left(\frac{\partial^2 u}{\partial\xi_i\partial\xi_j}\right) = (u_{ij})$. From (5.6.21) we have

$$\epsilon\rho^k(\lambda_1+\lambda_2) \le \left[\tfrac{4m\ell(\ell+1)\delta^2}{(\delta^2-\theta)^{\ell+2}} + \tfrac{2m\ell}{(\delta^2-\theta)^{\ell+1}}\right]\left(\tfrac{1}{\lambda_1}+\tfrac{1}{\lambda_2}\right) + \tfrac{8\epsilon}{k}\tfrac{m\ell}{(\delta^2-\theta)^{\ell+1}}d(\Omega^*)d(\Omega).$$

Namely

$$\epsilon\rho^{k+4} \le \tfrac{4\ell(\ell+2)md^2(\Omega^*)}{(\delta^2-\theta)^{\ell+2}} + \tfrac{8\epsilon}{k}\tfrac{m\ell}{(\delta^2-\theta)^{\ell+1}}d(\Omega^*)d(\Omega)\rho^2.$$

Using Young's inequality we have

$$\tfrac{8\epsilon}{k}\tfrac{m\ell}{(\delta^2-\theta)^{\ell+1}}d(\Omega^*)d(\Omega)\rho^2 \le \tfrac{\epsilon}{2}\rho^{k+4} + \left[\tfrac{8}{k}\tfrac{m\ell}{(\delta^2-\theta)^{\ell+1}}d(\Omega^*)d(\Omega)\right]^{\frac{k+4}{k+2}}.$$

It follows that

$$\rho^{k+4}\left(1+\epsilon\sum(x^k)^2\right)^{\frac{k+4}{k}} \le \tfrac{m^{\frac{k+4}{k+2}}C}{(\delta^2-\theta)^{\ell+2}},$$

where we choose $\ell = \frac{k}{2}$, and $C>1$ is a constant depending only on k, ℓ and $d(\Omega^*)$, $d(\Omega)$. Then

$$\rho^k\left(1+\epsilon\sum(x^k)^2\right) \le \frac{m^{\frac{k}{k+2}}C^{\frac{k}{k+4}}}{(\delta^2-\theta)^{\frac{(\ell+2)k}{k+4}}}.$$

Choose $m = C^{\frac{k(k+2)}{2(k+4)}}$, then

$$-\tfrac{m}{(\delta^2-\theta)^\ell} + \rho^k\left(1+\epsilon\sum(x^k)^2\right) \le 0.$$

It follows that

$$\rho^k \le m\left(\tfrac{4}{\delta^2}\right)^\ell$$

in $B^*_{\frac{\delta}{2}}(0)$. We use a covering argument to complete the proof of Lemma 5.6.6. ■

Using the same method we can prove

Lemma 5.6.7. *Let f be a smooth and strictly convex function defined on a bounded convex domain $\Omega \subset \mathbb{R}^2$ satisfying the PDE (4.5.9). Let Ω' be an arbitrary subdomain of Ω with dist$(\Omega', \partial\Omega) > 0$. Denote by Ω^* the Legendre transformation domain of f. Then the following estimate holds:*

$$\det(f_{ij}) \le b_0 \quad for \quad x \in \Omega',$$

where b_0 is a constant depending only on dist$(\Omega', \partial\Omega)$, diam$(\Omega)$, diam$(\Omega^*)$ and β.

Proof. Let $\dot{x} = (\dot{x}^1, \dot{x}^2) \in \Omega$ be an arbitrary point. By a translation and subtracting a linear function, we may assume that $\dot{x} = (0,0)$, $f(\dot{x}) = 0$ and grad $f(\dot{x}) = 0$. Choose $\delta > 0$ such that

$$0 < \delta < \operatorname{dist}(\dot{x}, \partial\Omega).$$

Consider the function

$$F := -\tfrac{m}{(\delta^2 - J^\sharp)^\ell} + \tfrac{1}{\rho^k}(1+\epsilon\theta)$$

defined on $B_\delta(0)$, where

$$\theta := \sum\left(\xi_j\right)^2, \quad J^\sharp := \sum\left(x^j\right)^2,$$

and where m, k, ℓ and ϵ are real positive constants to be determined later. Clearly, F attains its supremum at some interior point $\hat{x}$ of $B_\delta(0)$. We choose a local orthonormal frame field e_1, e_2 of the metric $\mathfrak{H}$ on M near $\hat{x}$. Then, at $\hat{x}$,

$$\begin{aligned} &-\tfrac{4m\ell(\ell+1)\delta^2}{(\delta^2-J^\sharp)^{\ell+2}}\sum f^{ii} - \tfrac{2m\ell}{(\delta^2-J^\sharp)^{\ell+1}}\sum f^{ii} - \tfrac{\epsilon}{k}\tfrac{8m\ell d(\Omega)d(\Omega^*)}{(\delta^2-J^\sharp)^{\ell+1}} + 2\epsilon\tfrac{1}{\rho^k}\sum f_{ii} \\ &+ k\left((k+1)+\beta\right)\tfrac{1}{\rho^{k+2}}(1+\epsilon\theta)\sum(\rho_{,i})^2 - 2\epsilon\,(k+1)\,\tfrac{1}{\rho^{k+1}}\sum\rho_{,i}\theta_{,i} \le 0. \end{aligned}$$

We choose a positive number k, depending only on β, such that

$$k(k+1+\beta) \ge 1.$$

It follows that

$$\begin{aligned} -\tfrac{2m\ell}{(\delta^2-J^\sharp)^{\ell+1}}\left(\tfrac{2(\ell+1)\delta^2}{\delta^2-J^\sharp}+1\right)\sum f^{ii} - \tfrac{\epsilon}{k}\tfrac{8m\ell}{(\delta^2-J^\sharp)^{\ell+1}}\,d(\Omega)\,d(\Omega^*) \\ + \tfrac{2\epsilon}{\rho^k}\left(1-2\epsilon\,(k+1)^2\,d(\Omega^*)^2\right)\sum f_{ii} \le 0. \end{aligned}$$

Choose $\epsilon > 0$ such that $4\epsilon(k+1)^2\mathrm{diam}(\Omega^*)^2 \leq 1$. Then we get

$$\tfrac{1}{\rho^k}(1+\epsilon\theta) \leq \frac{m^{\frac{k}{k+2}}}{(\delta^2 - J^\sharp)^{\frac{(\ell+2)k}{k+4}}} C^{\frac{k}{k+4}},$$

where $C > 1$ is a constant depending only on k, ℓ and $\mathrm{diam}(\Omega^*)$, $\mathrm{diam}(\Omega)$. Choose $m = C^{\frac{k(k+2)}{2(k+4)}}$, $\ell = \frac{k}{2}$. We obtain

$$-\tfrac{m}{(\delta^2 - J^\sharp)^\ell} + \tfrac{1}{\rho^k}(1+\epsilon\theta) \leq 0,$$

and Lemma 5.6.7 follows. ∎

Similar to Lemma 4.5.4, we obtain the following Lemma (for details see [55]):

Lemma 5.6.8. *Let $\Omega_k \subset \mathbb{R}^n$ be a sequence of smooth normalized convex domains, $f^{(k)}$ be a sequence of strictly convex C^∞ functions defined on Ω_k. Assume that*

$$\inf_{\Omega_k} f^{(k)} = f^k(q^k) = 0, \qquad f^{(k)} = C > 0 \quad on \ \partial\Omega_k.$$

Then there are constants $d > C$, $b > 0$ independent of k such that

$$\frac{\sum_i \left(\frac{\partial f^{(k)}}{\partial x^i}\right)^2}{(d+u^{(k)})^2} \leq b, \qquad k = 1, 2, \ldots \quad on \ \bar{\Omega}_k,$$

where $u^{(k)}$ is the Legendre transformation function of $f^{(k)}$ relative to 0.

Lemma 5.6.9. *Let Ω be a normalized convex domain. Assume that $f \in \mathcal{S}(\Omega, C)$ and f satisfies the PDE (4.5.9) with $\beta > 0$. Assume that u is the Legendre transformation function of f relative to 0. Then the following estimate holds:*

$$\tfrac{\det(f_{ij})}{(d+u)^{n+2}} \leq b_0 \qquad for \ x \in S_f(C') = \{x \in \Omega \mid f(x) \leq C'\},$$

where $C' < C$, and b_0 is a constant depending only on $\frac{C'}{C}$, C, and β.

Proof. Recall the definition of ρ and Φ from section 1.4. Consider the function

$$F := \exp\left\{-\tfrac{m}{C-f} + P\right\} \tfrac{1}{\rho(d+u)}$$

defined on the level set $S_f(C)$, where

$$P := \epsilon \tfrac{\sum(\xi_k)^2}{(d+u)^2},$$

$m = 4C$ and ϵ is a positive constant to be determined later. From Lemma 5.6.8 we know that P has an upper bound. Clearly, F attains its supremum at some interior point p^*. g, g' are defined as in Lemma 5.6.3. Near p^* we choose a local orthonormal frame field of the Calabi metric $\mathfrak{H}$. Then, at p^*,

$$-gf_{,i} + P_{,i} - \tfrac{\rho_{,i}}{\rho} - \tfrac{u_{,i}}{d+u} = 0,$$

$$-g'\sum(f_{,i})^2 - 2g - 2g\tfrac{\sum \rho_{,i}f_{,i}}{\rho} + (1+\beta)\Phi - \tfrac{2}{d+u} + 2\tfrac{\sum \rho_{,i}u_{,i}}{\rho(d+u)} + \tfrac{\sum(u_{,i})^2}{(d+u)^2} + \Delta P \leq 0.$$

Using the inequality of Schwarz we get

$$-\left(g' + \tfrac{4}{\beta}g^2\right)\sum (f_{,i})^2 + \tfrac{\beta}{4}\Phi + \tfrac{\beta}{2+\beta}\tfrac{\sum (u_{,i})^2}{(d+u)^2} + \Delta P - 2g - \tfrac{2}{d+u} \leq 0. \tag{5.6.22}$$

Now we calculate ΔP. By (3.3.7)

$$\sum (P_{,i})^2 \leq 8\epsilon P \tfrac{\sum u^{ii}}{(d+u)^2} + 8P^2 \tfrac{\sum (u_{,i})^2}{(d+u)^2}, \tag{5.6.23}$$

and

$$\begin{aligned}\Delta P &= \epsilon \tfrac{\Delta(\sum (\xi_k)^2)}{(d+u)^2} - 4\epsilon \tfrac{\mathfrak{H}(\text{grad} \sum (\xi_k)^2,\, \text{grad}\, u)}{(d+u)^3} - 2\epsilon \tfrac{\sum (\xi_k)^2 \Delta u}{(d+u)^3} + 6\epsilon \tfrac{\sum (\xi_k)^2 \sum (u_{,i})^2}{(d+u)^4} \\ &= \tfrac{\epsilon}{(d+u)^2}\left[2\sum u^{ii} - 2\mathfrak{H}(\text{grad} \ln \rho,\, \text{grad}\,(\sum (\xi_k)^2)) - 4\tfrac{\mathfrak{H}(\text{grad} \sum (\xi_k)^2,\, \text{grad}\, u)}{d+u}\right] \\ &\quad + 6\epsilon \tfrac{\sum (\xi_k)^2 \sum (u_{,i})^2}{(d+u)^4} + 4\epsilon \tfrac{\sum (\xi_k)^2 \mathfrak{H}(\text{grad} \ln \rho,\, \text{grad}\, u)}{(d+u)^3} - 4\epsilon \tfrac{\sum (\xi_k)^2}{(d+u)^3}.\end{aligned}$$

Note that

$$\|\text{grad}\left(\sum (\xi_i)^2\right)\|^2 = 4\sum u^{ij}\xi_i\xi_j.$$

Using the inequality of Schwarz we get

$$\Delta P \geq \tfrac{\epsilon}{2(d+u)^2}\sum u^{ii} - 18\epsilon b\Phi - 60\epsilon b \tfrac{\sum (u_{,i})^2}{(d+u)^2} - 4\epsilon b. \tag{5.6.24}$$

We choose ϵ such that

$$\epsilon b \leq \min\left\{\tfrac{\beta}{72},\ \tfrac{\beta}{60(2+\beta)},\ 1\right\}.$$

Insert (5.6.24) into (5.6.22), then

$$\tfrac{\epsilon}{2(d+u)^2}\sum u^{ii} - \left(g' + \tfrac{4}{\beta}g^2\right)\sum (f_{,i})^2 - 2g - 6 \leq 0.$$

Note that

$$\sum (f_{,i})^2 = \sum f^{ij}\xi_i\xi_j \leq b(d+u)^2 \sum f^{ii},$$

where b is the constant in Lemma 5.6.8. Denote by μ_1, μ_2 the eigenvalues of (u^{ij}), we have

$$\tfrac{\epsilon}{2(d+u)^2}(\mu_1 + \mu_2) - \left(g' + \tfrac{4}{\beta}g^2\right) b(d+u)^2 \left(\tfrac{1}{\mu_1} + \tfrac{1}{\mu_2}\right) - 2g - 6 \leq 0.$$

Thus

$$\tfrac{\epsilon}{2\rho^4(d+u)^4} - (g+3)\tfrac{1}{\rho^2(d+u)^2} - \left(g' + \tfrac{4}{\beta}g^2\right) b \leq 0,$$

where we used the inequality $\mu_1 + \mu_2 \geq 2\sqrt{\mu_1\mu_2}$. It follows that

$$\exp\left\{-\tfrac{m}{C-f} + P\right\} \tfrac{1}{\rho(d+u)} \leq b_1$$

for some constant b_1 depending on C and β. ∎

5.6.3 *Estimates for the third order derivatives*

In the following we derive estimates for the third order derivatives, this is the core of the proof of Theorem 5.6.2. We still restrict to functions of two variables. Let Ω be a normalized domain and $f \in \mathcal{S}(\Omega, C)$. Without loss of generality we assume that $\Phi \not\equiv 0$. We introduce the following notations:

$$\mathcal{A} := \max_{\Omega} \left\{ \exp\left\{ -\tfrac{m}{C-f} \right\} \tfrac{\Phi}{\rho^{\alpha}(d+u)^{\alpha}} \right\},$$

$$\mathcal{D} := \max_{\Omega} \left\{ \exp\left\{ -\tfrac{m}{C-f} + K \right\} \tfrac{g^2 \|\text{grad } f\|^2}{\rho^{\alpha}(d+u)^{\alpha}} \right\},$$

where

$$K := \tfrac{N}{\mathcal{A}} \exp\left\{ -\tfrac{m}{C-f} \right\} \tfrac{\Phi}{\rho^{\alpha}(d+u)^{\alpha}},$$

and m, α and N are positive constants to be determined later.

Lemma 5.6.10. *Assume that $\Omega \subset \mathbb{R}^2$ is a normalized domain and $f \in \mathcal{S}(\Omega, C)$ satisfies the PDE (4.5.9) with $\beta > 0$, and that there exists a constant $b > 0$ such that in Ω:*

$$\tfrac{1}{\rho(d+u)} < b.$$

Then there are constants $\alpha > 0$, N and m such that

$$\mathcal{A} \leq \max\{d_1, \tfrac{4}{\alpha\beta}\mathcal{D}\},$$

where d_1 is a constant depending only on α, C, b and β.

Proof. To prove this lemma we consider the function

$$F := \exp\left\{ -\tfrac{m}{C-f} \right\} W$$

defined on Ω, where

$$W := \tfrac{\Phi}{\rho^{\alpha}(d+u)^{\alpha}}.$$

Clearly, F attains its supremum at some interior point p^* in Ω. We may assume that $\Phi(p^*) \geq 1$, otherwise the proof is complete. Choose a local orthonormal frame field on M with respect to the Calabi metric. Then, at p^*,

$$\tfrac{W_{,i}}{W} - g f_{,i} = 0,$$

$$\tfrac{\Delta W}{W} - \tfrac{\|\text{grad } W\|^2}{W^2} - g' \|\text{grad } f\|^2 - g\Delta f \leq 0.$$

A direct calculation gives

$$W_{,i} = W \left[\tfrac{\Phi_{,i}}{\Phi} - \alpha \tfrac{\rho_{,i}}{\rho} - \alpha \tfrac{u_{,i}}{d+u} \right].$$

From Proposition 4.5.2, we have

$$\Delta\Phi \geq \tfrac{\|\text{grad } \Phi\|^2}{\Phi} + (2\beta^2 + 4\beta + 1)\Phi^2.$$

Thus we get

$$\begin{aligned}\Delta W =&\frac{\sum (W_{,i})^2}{W} + W\left[\frac{\Delta\Phi}{\Phi} - \frac{\sum(\Phi_{,i})^2}{\Phi^2} + \beta\alpha\Phi + \alpha\sum\left(\frac{\rho_{,i}}{\rho} + \frac{u_{,i}}{d+u}\right)^2 - \frac{2\alpha}{d+u}\right]\\ \geq&\frac{\sum (W_{,i})^2}{W} + W\left[(\beta\alpha + 2\beta^2 + 4\beta + 1)\Phi + \alpha\sum\left(\frac{\rho_{,i}}{\rho} + \frac{u_{,i}}{d+u}\right)^2 - \frac{2\alpha}{d+u}\right]\\ \geq&\frac{\sum (W_{,i})^2}{W} + W\left[(\beta\alpha + 2\beta^2 + 4\beta + 1)\Phi - 2\alpha\right].\end{aligned}$$

We choose $m = 4C$, then $2g' \leq g^2$. Using the inequality of Schwarz we get

$$\left[\beta\alpha + 2\beta^2 + 4\beta\right]\Phi - 2g^2\|\text{grad } f\|^2 - 2\alpha - 2g \leq 0.$$

We discuss two cases:

Case (i). If, at p^*,

$$2g^2\|\text{grad } f\|^2 \leq 2\alpha + 2g,$$

then,

$$\exp\left\{-\frac{m}{C-f}\right\}\frac{\Phi}{\rho^\alpha(d+u)^\alpha} \leq d'$$

for some constant $d' > 0$ depending only on α, β, b and C.

Case (ii). In the following we assume that

$$2g^2\|\text{grad } f\|^2(p^*) \geq 2\alpha + 2g.$$

Then we have

$$\alpha\beta\Phi \leq 4g^2\|\text{grad } f\|^2.$$

Multiply both sides with $\exp\{-\frac{m}{C-f} + K\}\frac{1}{\rho^\alpha(d+u)^\alpha}(p^*)$, we have

$$\exp\{N\}\mathcal{A} \leq \frac{4}{\alpha\beta}\exp\left\{-\frac{m}{C-f} + N\right\}\frac{g^2\|\text{grad } f\|^2}{\rho^\alpha(d+u)^\alpha}(p^*).$$

Note that $K(p^*) = N$. Hence

$$\exp\{N\}\mathcal{A} \leq \frac{4}{\alpha\beta}\mathcal{D}.$$

In (i) and (ii) we got upper estimates, thus Lemma 5.6.10 is proved. ■

Lemma 5.6.11. *Let $\Omega \subset \mathbb{R}^2$ be a normalized convex domain, $f \in \mathcal{S}(\Omega, C)$ satisfying the PDE (4.5.9) with $\beta > 0$. Assume that there exists a constant $b > 0$ such that in Ω:*

$$\frac{1}{\rho(d+u)} < b.$$

Then there exist constants $\alpha > 0$, N and m such that

$$\mathcal{A} \leq d_2, \qquad \mathcal{D} \leq d_2$$

for some constant $d_2 > 0$ depending only on α, b, β and C.

Proof. Without loss of generality we may assume that $\frac{4}{\alpha\beta}\mathcal{D} \geq d_1$. We put

$$\tau^\sharp := K, \quad Q := g^2\frac{1}{\rho^\alpha(d+u)^\alpha}, \quad h := f$$

in (5.6.6). Suppose that F attains its supremum at the point q^*. Choose a local orthonormal frame field with respect to the Calabi metric $\mathfrak{H}$ near q^*. From $F_{,i}(q^*) = 0$ we have, at q^*,

$$\left(-gf_{,i} + \tfrac{4}{C-f}f_{,i} - \alpha\tfrac{\rho_{,i}}{\rho} - \alpha\tfrac{u_{,i}}{d+u} + K_{,i}\right)(f_{,1})^2 + 2\sum f_{,j}f_{,ji} = 0. \qquad (5.6.25)$$

We insert (5.6.11) into (5.6.7) and choose $\delta = \frac{1}{10}$, and get, at q^*,

$$\begin{aligned} 2(f_{,11})^2 + 2(f_{,12})^2 + 4\tfrac{\rho_{,11}}{\rho}(f_{,1})^2 + (\beta\alpha - 388)\Phi(f_{,1})^2 - 328 + \Delta K(f_{,1})^2 \\ - \left(g'(f_{,1})^2 + 2\left(g - \tfrac{4}{C-f}\right) + 2\left(g - \tfrac{4}{C-f}\right)\tfrac{\rho_{,1}}{\rho}f_{,1}\right. \\ \left. -\alpha\sum\left(\tfrac{\rho_{,i}}{\rho} + \tfrac{u_{,i}}{d+u}\right)^2 + \tfrac{2\alpha}{d+u}\right)(f_{,1})^2 \le 0. \qquad (5.6.26) \end{aligned}$$

Consider the inequality (5.6.26). In the following we choose $N >> 10$ and calculate estimates for the terms $(f_{,11})^2 + (f_{,12})^2$, ΔK and $4\frac{\rho_{,11}}{\rho}(f_{,1})^2$, respectively.

(1) Estimate for $(f_{,11})^2 + (f_{,12})^2$.

From (5.6.25) we have

$$\begin{aligned} 2(f_{,11})^2 &= \tfrac{1}{2}\left[gf_{,1} - \tfrac{4}{C-f}f_{,1} + \alpha\left(\tfrac{\rho_{,1}}{\rho} + \tfrac{u_{,1}}{d+u}\right) - K_{,1}\right]^2(f_{,1})^2 \\ &\ge \tfrac{3}{4N}\left[\left(g - \tfrac{4}{C-f}\right)f_{,1} + \alpha\left(\tfrac{\rho_{,1}}{\rho} + \tfrac{u_{,1}}{d+u}\right)\right]^2(f_{,1})^2 - \tfrac{9}{10}\tfrac{(K_{,1})^2}{K}(f_{,1})^2, \qquad (5.6.27) \end{aligned}$$

where we used the fact $K \le N$ and the elementary inequality

$$(a+b)^2 \ge (1-\delta)a^2 - (\tfrac{1}{\delta} - 1)b^2$$

with $a := \left((g - \frac{4}{C-f})f_{,1} + \alpha\left(\frac{\rho_{,1}}{\rho} + \frac{u_{,1}}{d+u}\right)\right)$, $b := K_{,1}$, and $\delta := \frac{10K}{18+10K}$. Similarly

$$2(f_{,12})^2 \ge \tfrac{3}{4N}\alpha^2\left(\tfrac{\rho_{,2}}{\rho} + \tfrac{u_{,2}}{d+u}\right)^2(f_{,1})^2 - \tfrac{9}{10}\tfrac{(K_{,2})^2}{K}(f_{,1})^2. \qquad (5.6.28)$$

(2) Estimate for ΔK.

$$K_{,i} = K\left(\tfrac{\Phi_{,i}}{\Phi} - \alpha\tfrac{\rho_{,i}}{\rho} - \alpha\tfrac{u_{,i}}{d+u} - gf_{,i}\right), \qquad (5.6.29)$$

$$\begin{aligned} \Delta K =& \tfrac{\sum(K_{,i})^2}{K} + K\left((\beta\alpha + 2\beta^2 + 4\beta + 1)\Phi + \alpha\sum\left(\tfrac{\rho_{,i}}{\rho} + \tfrac{u_{,i}}{d+u}\right)^2\right) \\ &- K\left(g'(f_{,1})^2 + 2g + \tfrac{2\alpha}{d+u} + 2g\tfrac{\rho_{,1}}{\rho}f_{,1}\right). \qquad (5.6.30) \end{aligned}$$

Then we get

$$\Delta K \ge \tfrac{\|\text{grad } K\|^2}{K} - 2Kg\,\tfrac{\rho_{,1}f_{,1}}{\rho} - N\left(g'(f_{,1})^2 + 2g + \tfrac{2\alpha}{d+u}\right). \qquad (5.6.31)$$

(3) Estimate for $4\frac{\rho_{,11}}{\rho}(f_{,1})^2$.

First, we choose a frame $\tilde{e}_1, \tilde{e}_2$ such that $\rho_{,1} = \|\text{grad}\ \rho\|$. Note that

$$\Phi_{,i} = \frac{2\rho_{,1}\rho_{,1i}}{\rho^2} - 2\frac{\rho_{,i}(\rho_{,1})^2}{\rho^3}.$$

It is easy to check

$$\frac{(\rho_{,11})^2+(\rho_{,12})^2}{\rho^2} \le \frac{\sum(\Phi_{,i})^2}{2\Phi} + 2\Phi^2.$$

Then

$$\begin{aligned}\sum\frac{(\rho_{,ij})^2}{\rho^2} &= \frac{(\rho_{,11})^2+2(\rho_{,12})^2+\left(\frac{\beta(\rho_{,1})^2}{\rho}+\rho_{,11}\right)^2}{\rho^2}\\ &\le \frac{3((\rho_{,11})^2+(\rho_{,12})^2)}{\rho^2} + 2\beta^2\Phi^2 \le \frac{3\sum(\Phi_{,i})^2}{2\Phi} + (6+2\beta^2)\Phi^2.\end{aligned}$$

Next, we return to the frame e_1, e_2. It follows that

$$\begin{aligned}4\frac{|\rho_{,11}|}{\rho}(f_{,1})^2 \le& 2\sqrt{6}\ \frac{\|\text{grad}\ \Phi\|}{\sqrt{\Phi}}(f_{,1})^2 + 4\sqrt{6+2\beta^2}\ \Phi(f_{,1})^2\\ \le& 4\sqrt{3}\sqrt{\Phi}\left(\left[\sum\left(\frac{\Phi_{,i}}{\Phi} - gf_{,i} - \alpha\left(\frac{\rho_{,i}}{\rho} + \frac{u_{,i}}{d+u}\right)\right)^2\right]^{1/2}\right.\\ &+\left.\left[\sum\left(gf_{,i} + \alpha\left(\frac{\rho_{,i}}{\rho} + \frac{u_{,i}}{d+u}\right)\right)^2\right]^{1/2}\right)(f_{,1})^2 + (12+8\beta)\Phi(f_{,1})^2.\end{aligned} \tag{5.6.32}$$

We apply the inequality of Schwarz and (5.6.29) and get

$$\begin{aligned}&4\sqrt{3}\sqrt{\Phi}\left[\sum\left(\frac{\Phi_{,i}}{\Phi} - gf_{,i} - \alpha\left(\frac{\rho_{,i}}{\rho} + \frac{u_{,i}}{d+u}\right)\right)^2\right]^{1/2}(f_{,1})^2\\ &\quad\le \frac{1}{20}\frac{\|\text{grad}\ K\|^2}{K}(f_{,1})^2 + \frac{240}{N}\mathcal{A}\exp\left\{\frac{m}{C-f}\right\}\rho^\alpha(d+u)^\alpha(f_{,1})^2,\end{aligned} \tag{5.6.33}$$

$$\begin{aligned}&4\sqrt{3}\sqrt{\Phi}\left[\sum\left(gf_{,i} + \alpha\left(\frac{\rho_{,i}}{\rho} + \frac{u_{,i}}{d+u}\right)\right)^2\right]^{1/2}(f_{,1})^2 \le 300N\Phi(f_{,1})^2\\ &\quad+ \frac{1}{12N}\sum\left[\left(g - \frac{4}{C-f}\right)f_{,i} + \alpha\left(\frac{\rho_{,i}}{\rho} + \frac{u_{,i}}{d+u}\right)\right]^2(f_{,1})^2 + \frac{4}{3N}\frac{1}{(C-f)^2}(f_{,1})^4.\end{aligned} \tag{5.6.34}$$

From Lemma 5.6.10, note that $K(p^*) = N$, then

$$\exp\{N\}\mathcal{A} \le \frac{4}{\alpha\beta}\exp\left\{-\frac{m}{C-f} + K\right\}\frac{g^2\|\text{grad}\ f\|^2}{\rho^\alpha(d+u)^\alpha}(q^*).$$

It follows that, at q^*,

$$\frac{240}{N}\mathcal{A}\exp\left\{\frac{m}{C-f}\right\}\rho^\alpha(d+u)^\alpha(f_{,1})^2 \le \frac{960}{N\alpha\beta}\ g^2(f_{,1})^4. \tag{5.6.35}$$

We inserting (5.6.33), (5.6.34) and (5.6.35) into (5.6.32) and obtain

$$\begin{aligned}4\frac{|\rho_{,11}|}{\rho}(f_{,1})^2 \le& \frac{1}{20}\frac{\|\text{grad}\ K\|^2}{K}(f_{,1})^2 + \frac{960}{N\alpha\beta}g^2(f_{,1})^4\\ &+ \frac{4}{3N}\frac{1}{(C-f)^2}(f_{,1})^4 + (302N+8\beta)\Phi(f_{,1})^2\\ &+ \frac{1}{12N}\sum\left[\left(g - \frac{4}{C-f}\right)f_{,i} + \alpha\left(\frac{\rho_{,i}}{\rho} + \frac{u_{,i}}{d+u}\right)\right]^2(f_{,1})^2.\end{aligned} \tag{5.6.36}$$

We insert (5.6.27), (5.6.28), (5.6.31) and (5.6.36) into (5.6.26), this gives

$$\frac{2}{3N}\sum\left[\left(g-\frac{4}{C-f}\right)f_{,i}+\alpha\left(\frac{\rho_{,i}}{\rho}+\frac{u_{,i}}{d+u}\right)\right]^2(f_{,1})^2+\alpha\sum\left(\frac{\rho_{,i}}{\rho}+\frac{u_{,i}}{d+u}\right)^2(f_{,1})^2$$
$$+(\alpha\beta-340N-8\beta)\Phi(f_{,1})^2-2\left(Kg+g-\frac{4}{C-f}\right)\frac{\rho_{,1}}{\rho}(f_{,1})^3-328$$
$$-\left[(N+1)g'+\frac{1000}{N\alpha\beta}g^2+\frac{4}{3N}\frac{1}{(C-f)^2}\right](f_{,1})^4-2(N+1)\left(g+\frac{\alpha}{d+u}\right)(f_{,1})^2\leq 0. \tag{5.6.37}$$

We choose N and α such that

$$1+N=\frac{2\alpha}{3N},\quad i.e.,\quad \alpha=\frac{3N(1+N)}{2}, \tag{5.6.38}$$

and choose N large enough so that

$$\beta\alpha-340N-8\beta=\left(\frac{3N(1+N)}{2}-8\right)\beta-340N>200N.$$

We choose $m\geq 2C\alpha\beta N(N+1)$, then

$$g'(N+1)\leq\frac{1}{N\alpha\beta}g^2,\qquad \frac{4}{3N}\frac{1}{(C-f)^2}<\frac{1}{N^2\alpha\beta}g^2.$$

In the following we discuss two cases:

Case 1: $\sum\frac{\rho_{,i}f_{,i}}{\rho}>0$. In this case, by (5.6.38), we have

$$\frac{2}{3N}\left[\left(g-\frac{4}{C-f}\right)f_{,i}+\alpha\left(\frac{\rho_{,i}}{\rho}+\frac{u_{,i}}{d+u}\right)\right]^2(f_{,1})^2-(2+2N)\left(g-\frac{4}{C-f}\right)\frac{\rho_{,1}}{\rho}(f_{,1})^3$$
$$\geq\frac{2}{3N}\left(g-\frac{4}{C-f}\right)^2(f_{,1})^4-2(N+1)\left(g-\frac{4}{C-f}\right)(f_{,1})^2\geq\frac{1}{3N}g^2(f_{,1})^4-C_0,$$

where we used the fact $\frac{|\sum f_{,i}u_{,i}|}{d+u}=\frac{|f+u|}{d+u}<1$ (see (4.5.28)); here C_0 is a constant depending only on β. Note that $K\leq N$, we have

$$2N\left(g-\frac{4}{C-f}\right)\frac{\rho_{,1}f_{,1}}{\rho}-2gK\frac{\rho_{,1}f_{,1}}{\rho}\geq-\frac{8N}{C-f}\frac{\rho_{,1}f_{,1}}{\rho}\geq-200N\Phi-\frac{3}{50\alpha\beta}g^2(f_{,1})^2.$$

Then, we have

$$\frac{1}{6N}g^2(f_{,1})^4-2(N+1)\left(g+\frac{\alpha}{d+u}\right)(f_{,1})^2-C_1\leq 0$$

for some constant C_1 depending only on β.

Case 2: $\sum\frac{\rho_{,i}f_{,i}}{\rho}\leq 0$. By the inequality of Schwarz we have

$$\frac{2}{3N}\sum\left[\left(g-\frac{4}{C-f}\right)f_{,i}+\alpha\left(\frac{\rho_{,i}}{\rho}+\frac{u_{,i}}{d+u}\right)\right]^2(f_{,1})^2+\alpha\sum\left(\frac{\rho_{,i}}{\rho}+\frac{u_{,i}}{d+u}\right)^2(f_{,1})^2$$
$$\geq\frac{2}{2\alpha+3N}\left(g-\frac{4}{C-f}\right)^2(f_{,1})^4\geq\frac{1}{2\alpha+3N}g^2(f_{,1})^4.$$

Then

$$\frac{1}{4\alpha+6N}g^2(f_{,1})^4-2(N+1)\left(g+\frac{\alpha}{d+u}\right)(f_{,1})^2-328\leq 0.$$

Consequence: In both cases, we have an inequality of the type

$$a_0 g^2 (f_{,1})^4 - (a_1 g + a_2)(f_{,1})^2 - a_3 \leq 0,$$

where a_0, a_1, a_2 and a_3 are positive constants depending only on β. Consequently

$$\exp\left\{-\frac{m}{C-f} + K\right\} g^2 \frac{1}{\rho^\alpha (d+u)^\alpha} \|\text{grad } f\|^2 \leq a_4.$$

Together with Lemma 5.6.10, Lemma 5.6.11 follows. ■

As a corollary of Lemma 5.6.11, we get the following estimates:

Proposition 5.6.12. *Let $\Omega \subset \mathbb{R}^2$ be a normalized convex domain and $0 \in \Omega$ be the center of Ω. Let f be a strictly convex C^∞ function defined on Ω. Assume that*

$$\inf_\Omega f = 0, \qquad f = C > 0 \quad on \ \partial\Omega$$

and that f satisfies the PDE (4.5.9) with $\beta > 0$. Then there exists a constant $\alpha > 0$ such that, on $\Omega_{\frac{C}{2}} := \{x \in \Omega \mid f(x) < \frac{C}{2}\}$,

$$\frac{\Phi}{\rho^\alpha (d+u)^\alpha} \leq d_3, \quad \frac{\|grad\ f\|^2}{\rho^\alpha (d+u)^\alpha} \leq d_3$$

for some constant $d_3 > 0$ depending only on β and C.

Proposition 5.6.13. *Let $\Omega \subset \mathbb{R}^2$ be a normalized convex domain and $0 \in \Omega$ be the center of Ω. Let f be a strictly convex C^∞ function defined on Ω. Assume that*

$$\inf_\Omega f = 0, \qquad f = C > 0 \quad on \ \partial\Omega$$

and that f satisfies the PDE (4.5.9) with $\beta = 0$, and that there is a constant $b > 0$ such that, in Ω,

$$\frac{1}{\rho} \leq b.$$

Then there exists $\alpha > 0$ such that the following estimates hold on $\Omega_{\frac{C}{2}}$:

$$\frac{\Phi}{\rho^\alpha} \leq d_4, \qquad \frac{\|grad\ f\|^2}{\rho^\alpha} \leq d_4$$

for some constant $d_4 > 0$ depending only on α, b and C.

Since the proof is very similar to the proof of Proposition 5.6.12, we omit it here.

5.6.4 *Estimates for* $\sum f_{ii}$

In the following we will derive an upper bound for $\sum f_{ii}$.

Proposition 5.6.14. *Let $\Omega \subset \mathbb{R}^2$ be a normalized convex domain. Let $f \in \mathcal{S}(\Omega, C)$ be a smooth and strictly convex function defined in Ω, which satisfies the equation (4.5.9) with $\beta > 0$. Assume that there are constants $d_3 > 0$ and $\alpha > 1$ such that, in Ω,*

$$\frac{\Phi}{\rho^\alpha (d+u)^\alpha} < d_3, \quad \frac{1}{\rho(d+u)} < d_3.$$

Then there exists a constant $d_5 > 0$, depending only on α, β, d_3 and C, such that

$$\exp\left\{-\tfrac{32(2+d_3)C}{C-f}\right\}\tfrac{\sum f_{ii}}{\rho^\alpha(d+u)^{\alpha+2}} \le d_5$$

on Ω.

Proof. Put

$$h := \xi_1, \quad \tau^\sharp := W + P, \quad Q := \tfrac{1}{\rho^\alpha(d+u)^{\alpha+2}}$$

in (5.6.6), where

$$P := \epsilon\tfrac{\sum(\xi_k)^2}{(d+u)^2}, \quad W := \tfrac{\Phi}{\rho^\alpha(d+u)^\alpha}.$$

We choose ϵ such that $P \le \frac{1}{30}$ on Ω. F attains its supremum at some point p^*. Choose a local orthonormal frame field on M with respect to the Calabi metric $\mathfrak{H}$ near p^*. From $F_{,i}(p^*) = 0$ we have, at p^*,

$$\left(-gf_{,i} - \alpha\tfrac{\rho_{,i}}{\rho} - (\alpha+2)\tfrac{u_{,i}}{d+u} + W_{,i} + P_{,i}\right)\sum(h_{,j})^2 + 2\sum h_{,j}h_{,ji} = 0. \qquad (5.6.39)$$

We insert (5.6.12) into (5.6.7) with $\delta := \frac{1}{12}$, and use the inequality of Schwarz, this gives, at p^*,

$$\begin{aligned}&3\sum(h_{,1i})^2 + (\Delta W + \Delta P)(h_{,1})^2 - 4\tfrac{\rho_{,11}}{\rho}(h_{,1})^2 - a_1\Phi(h_{,1})^2 + (\alpha+1)\tfrac{\sum(u_{,i})^2}{(d+u)^2}(h_{,1})^2\\ &-\left[\left(g' + \tfrac{1}{32(2+d_3)}g^2\right)\sum(f_{,i})^2 + 2g + 2(\alpha+2)\right](h_{,1})^2 \le 0, \qquad (5.6.40)\end{aligned}$$

where here and later we use a_i to denote constants depending only on α and β. We calculated ΔP in Section 5.6.2 (see (5.6.24)). Now we estimate ΔW. We use Propositions 4.5.2 and (5.6.12) with $\delta = \frac{1}{16\alpha}$ to obtain

$$\Delta W \ge \left(1 - \tfrac{1}{2\alpha}\right)\tfrac{\sum(W_{,i})^2}{W} + \tfrac{\sum(\rho_{,ij})^2}{8\alpha\rho^{\alpha+2}(d+u)^\alpha} - a_2\Phi - a_3. \qquad (5.6.41)$$

By the inequality of Schwarz we have

$$4\tfrac{|\rho_{,11}|}{\rho} \le \tfrac{\sum(\rho_{,ij})^2}{8\alpha\rho^{\alpha+2}(d+u)^\alpha} + 32\alpha\rho^\alpha(d+u)^\alpha. \qquad (5.6.42)$$

Now we calculate $\sum(h_{,1i})^2$. From (5.6.39) we get

$$\sum(h_{,1i})^2 = \tfrac{1}{4}\sum\left(gf_{,i} + \alpha\tfrac{\rho_{,i}}{\rho} + (\alpha+2)\tfrac{u_{,i}}{d+u} - W_{,i} - P_{,i}\right)^2(h_{,1})^2.$$

We estimate

$$\begin{aligned}&\tfrac{1}{4}\sum\left(gf_{,i} + \alpha\tfrac{\rho_{,i}}{\rho} + (\alpha+2)\tfrac{u_{,i}}{d+u} - W_{,i} - P_{,i}\right)^2\\ &\ge \tfrac{1}{8}\sum\left(gf_{,i} + (\alpha+2)\tfrac{u_{,i}}{d+u} - W_{,i}\right)^2 - \tfrac{1}{2}\sum(P_{,i})^2 - \tfrac{1}{2}\alpha^2\Phi\\ &\ge \tfrac{1}{8(1+W)}\sum\left(gf_{,i} + (\alpha+2)\tfrac{u_{,i}}{d+u}\right)^2 - \tfrac{\sum(W_{,i})^2}{8W} - \tfrac{1}{2}\sum(P_{,i})^2 - \tfrac{1}{2}\alpha^2\Phi\\ &\ge \tfrac{1}{8(1+d_3)}g^2\sum(f_{,i})^2 - (\alpha+2)g - \tfrac{\sum(W_{,i})^2}{8W} - \tfrac{1}{2}\sum(P_{,i})^2 - \tfrac{1}{2}\alpha^2\Phi. \qquad (5.6.43)\end{aligned}$$

In the second inequality above we used the inequality $(a+b)^2 \geq (1-\eta)a^2 - (\frac{1}{\eta}-1)b^2$ with $\eta = \frac{W}{1+W}$. In the third inequality above we used the fact $\frac{|\sum f_{,i}u_{,i}|}{d+u} = \frac{|f+u|}{d+u} < 1$.

We choose $m = 32(2+d_3)C$, then $\left(g' + \frac{1}{32(2+d_3)}g^2\right) < \frac{3}{32(2+d_3)}g^2$. Since

$$\left(\frac{\partial^2 u}{\partial \xi_i \partial \xi_j}\right) = \left(\frac{\partial^2 f}{\partial x^i \partial x^j}\right)^{-1} \quad \text{and} \quad (h_{,1})^2 = \sum f_{1i}f_{1j}f^{ij} = f_{11},$$

we have $\sum u^{ii} \geq (h_{,1})^2$. Choose ϵ as in Lemma 5.6.9. We insert (5.6.23), (5.6.24), (5.6.41), (5.6.42), (5.6.43) into (5.6.40) and obtain

$$\frac{\epsilon}{10(d+u)^2}(h_{,1})^2 - a_6\Phi - a_7 g - a_8\rho^\alpha(d+u)^\alpha - a_9 \leq 0.$$

It follows that

$$\exp\left\{-\frac{32(2+d_3)C}{C-f}\right\}\frac{\|\text{grad}\ \xi_1\|^2}{\rho^\alpha(d+u)^{\alpha+2}} \leq d_5$$

for some constant d_5. As $\|\text{grad}\ \xi_1\|^2 = f_{11}$, we can prove this inequality for any f_{ii} in the same way. This completes the proof. ■

In a similar way we can prove the following Proposition:

Proposition 5.6.15. *Let $x^3 = f(x^1, x^2)$ be a smooth and strictly convex function defined on a normalized convex domain $\Omega \subset \mathbb{R}^2$, which satisfies the equation (4.5.9) with $\beta = 0$. Assume that $f \in \mathcal{S}(\Omega, C)$, and that there exist constants $\alpha \geq 0$ and $d_4 \geq 0$ such that*

$$\frac{\Phi}{\rho^\alpha} \leq d_4, \quad \frac{1}{\rho} \leq d_4$$

on $\bar{\Omega}$. Then there is a constant $d_5 > 0$, depending only on α, d_4 and C, such that

$$\exp\left\{-\frac{32(2+d_4)C}{C-f}\right\}\frac{\sum f_{ii}}{\rho^\alpha(d+u)^{\alpha+2}} \leq d_5$$

on Ω.

Remark. It is easy to see that Propositions 5.6.14 and 5.6.15 hold for any dimension.

5.6.5 *Proof of Theorem 5.6.2*

We begin with the following Lemma:

Lemma 5.6.16. *Let $\Omega_k \subset \mathbb{R}^2$ be a sequence of smooth normalized convex domains, converging to a convex domain Ω, and let $f^{(k)} \in \mathcal{S}(\Omega_k, C)$ with $f^k(q^k) = 0$. Assume that the functions $f^{(k)}$ satisfy the PDE (4.5.9) with $\beta \geq 0$. Then there exists a subsequence $f^{(i_\ell)}$ that locally uniformly converges to a convex function $f \in C^0(\Omega)$ with $d(p_o, \partial\Omega) > 0$, where p_o is the point such that $f(p_o) = 0$. Moreover, there is an open neighborhood N of p_o such that $f^{(i_\ell)}$ converges to f, and also all their derivatives converge, therefore f is smooth and strictly convex in N.*

Proof. Case $\beta > 0$. Let $0 \in \Omega_k$ be the center of Ω_k and $u^{(k)}$ the Legendre transformation function of $f^{(k)}$ relative to 0.

To simplify the notations we will use $f^{(k)}$ to denote $f^{(i_\ell)}$. By Lemmas 5.6.8, 5.6.9 and Propositions 5.6.12, 5.6.14 we have the uniform estimates

$$\frac{\Phi^{(k)}}{\rho^\alpha (d+u^{(k)})^\alpha} \leq d_6, \quad \frac{1}{\rho^\alpha (d+u^{(k)})^\alpha} \leq d_6, \quad \frac{\sum f_{ii}^{(k)}}{\rho^\alpha (d+u^{(k)})^{\alpha+2}} \leq d_6$$

in

$$S_{f^{(k)}}(q_k, \tfrac{C}{2}) := \{x \in \Omega_k \mid f^{(k)} < \tfrac{C}{2}\},$$

where d_6 is a positive constant depending only on β and C. We may assume that q_k converges to p_o. Let $B_R(q_k)$ be a Euclidean ball such that $\Omega \subset B_{\frac{R}{2}}(q_k)$. Then the Legendre transformation domain of Ω satisfies that $B^*_\delta(0) \subset \Omega^*$, where $\delta = \frac{C}{2R}$ and $B^*_\delta(0) = \{\xi \mid \xi_1^2 + \xi_2^2 < \delta^2\}$. By Lemma 5.6.6, we have

$$\det(f_{ij}) \geq b_3$$

for $\xi \in B^*_{\frac{\delta}{2}}(0)$ where b_3 is constant depending only on C and β. Restricting to $B^*_\delta(0)$, we have

$$-\tfrac{C}{2R} - C \leq u^{(k)} = \sum \xi_i x^i - f^{(k)} \leq \tfrac{C}{2R}.$$

Therefore, the sequence $u^{(k)}$ locally uniformly converges to a convex function u^∞ in $B^*_\delta(0)$, and there are constants $0 < \lambda \leq \Lambda < \infty$ such that the following estimates hold in $B^*_{\frac{\delta}{2}}(0)$

$$\lambda \leq \lambda_i^{(k)} \leq \Lambda, \qquad \text{for} \quad i = 1, 2, \cdots, n, \quad k = 1, 2, \cdots$$

where $\lambda_1^{(k)}, \cdots, \lambda_n^{(k)}$ denote the eigenvalues of the matrix $(f_{ij}^{(k)})$. Then, by standard elliptic estimates, Lemma 5.6.16 follows in case $\beta > 0$.

Case $\beta = 0$. Denote $D := \{x \mid f(x) = 0\}$. Again we have two subcases.
(i) If $D \bigcap \partial\Omega = \emptyset$ then there is a constant $h > 0$ such that the level set satisfies $\bar{S}_f(p_o, h) \subset \Omega$, and so we have a uniform estimate for $\sum \left(\frac{\partial f^{(k)}}{\partial x^i}\right)^2$ in $\bar{S}_{f^{(k)}}(p_k, h)$. From Lemma 5.6.4, it follows that there is a uniform estimate for $\frac{1}{\rho}$ in $\bar{S}_f^{(k)}(p_k, \frac{h}{2})$. Then we use Propositions 5.6.13 and 5.6.15 and the same argument as above to complete the proof.
(ii) In case $D \bigcap \partial\Omega \neq \emptyset$, let $p \in D \bigcap \partial\Omega$. Since the PDE (4.5.9) with $\beta = 0$ is an equiaffine invariant, we may choose a new coordinate system such that $\sum \left(\frac{\partial f^{(k)}}{\partial x^i}\right)^2$ is uniformly bounded in $\bar{S}_{f^{(k)}}(p, h)$. Then the same argument shows that f is smooth in a neighborhood of p, and we get a contradiction. This excludes the case $D \bigcap \partial\Omega \neq \emptyset$ and thus completes the proof of this lemma. ■

Remark. Here we have shown that there is a uniform estimate

$$0 < b_4 < \det(u_{ij}^{(k)}) < b_5 < \infty$$

in $B^*_{\frac{\delta}{2}}(0)$ for some constants b_4, b_5. We can use the convex body theory and a theorem of Aleksandrov ([13] p.35) to conclude that u^∞ is strictly convex. Then we can also use the Caffarelli-Gutierrez theory to prove this lemma, for details see [53].

Proof of Theorem 5.6.2. Let $x : M \to \mathbb{R}^3$ be a locally strongly convex surface, given as graph of a smooth, strictly convex function f defined for all $(x^1, x^2) \in \mathbb{R}^2$. Assume that f satisfies the PDE (4.5.9) with $\beta \geq 0$. Given any $p \in M$, by adding a linear function, we may assume that

$$f(p) = 0, \quad \frac{\partial f}{\partial x^i}(p) = 0, \; i = 1, 2.$$

Choose a sequence $\{C_k\}$ of positive numbers such that $C_k \to \infty$ as $k \to \infty$. Then, for any C_k, the *section*

$$S_f(p, C_k) = \left\{(x^1, x^2) \in \mathbb{R}^2 \mid f(x^1, x^2) < C_k\right\}$$

is a bounded convex domain in $\mathbb{R}^2$. It is well-known that (see section 4.5.6) there exists a unique ellipsoid E_k which attains the minimum volume among all ellipsoids that contain $S_f(p, C_k)$ and that are centered at the center of mass of $S_f(p, C_k)$ such that

$$2^{-\frac{3}{2}} E_k \subset S_f(p, C_k) \subset E_k.$$

Let T_k be an affine transformation such that

$$T_k(E_k) = B_1(0) = \left\{(x^1, x^2) \in \mathbb{R}^2 \mid (x^1)^2 + (x^2)^2 < 1\right\}.$$

Define the functions

$$f^{(k)}(x) = \frac{f(T_k^{-1}x)}{C_k}.$$

Then

$$B_{2^{-\frac{3}{2}}}(0) \subset \Omega_k \subset B_1(0),$$

where

$$\Omega_k = \left\{(x^1, x^2) \in \mathbb{R}^2 \mid f^{(k)}(x^1, x^2) < 1\right\}.$$

Taking subsequences we may assume that $\{\Omega_k\}$ converges to a convex domain Ω and $\{f^{(k)}\}$ converges to a convex function f^∞, locally uniformly in Ω. By Lemma 5.6.16 the function f^∞ is smooth and strictly convex in a neighborhood of $T^k(p) \in \Omega$. It follows that the functions $\Phi^{(k)}(T^k(p))$ are uniformly bounded.

Assume that $\Phi(p) \neq 0$; by a direct calculation we have

$$\Phi^{(k)}(T^k(p)) = C_k \Phi(p) \to \infty,$$

thus we get a contradiction, and thus $\Phi(p) = 0$. Since p is arbitrary, we have $\Phi = 0$ everywhere on M. It follows that $\det(f_{ij}) = const$. So f must be a quadratic polynomial. This completes the proof of Theorem 5.6.2. ■

5.7 An Affine Bernstein Problem in Dimension 3

In this section we use the following *standard notation*: we consider a domain $\Omega \subset \mathbb{R}^n$, $n \geq 2$, and a locally strongly convex function $f : \Omega \to \mathbb{R}$. We consider the graph $x^{n+1} = f(x^1, ..., x^n)$ and consider the hypersurface $M = \{(x, f(x)) \mid x \in \Omega\}$. For a fixed point $p_0 \in M$ denote by $d(p_0, p)$ the geodesic distance from p to p_0 with respect to the Calabi metric $\mathfrak{H}$, and by $r : p \mapsto r(p_0, p) = d(p_0, p)$ the geodesic distance function. For any $0 < a \in \mathbb{R}$ let $\bar{B}_a(p_0, \mathfrak{H}) := \{p \in M \mid d(p_0, p) \leq a\}$ be a closed geodesic ball. From section 1.4 recall the definition of Φ, and define $F : \bar{B}_a(p_0, \mathfrak{H}) \mapsto \mathbb{R}$ by

$$F(p) := (a^2 - r^2(p_0, p))^2 \Phi(p).$$

Obviously, F attains its supremum at some interior point p^*. We may assume that r^2 is a C^2-function in a neighborhood of p^*, and $\Phi > 0$ at p^*.

We start this section with Theorem 5.7.1, and extend this result in section 5.8 below.

Theorem 5.7.1. [54]. *Let $x^{n+1} = f(x^1, \cdots, x^n)$ be a locally strongly convex function defined in a domain $\Omega \subset \mathbb{R}^n$. If $M = \{(x, f(x)) \mid x \in \Omega\}$ is an affine maximal hypersurface, and if M is complete with respect to the Calabi metric $\mathfrak{H}$, then, for dimension $n = 2$ or $n = 3, M$ must be an elliptic paraboloid.*

We divide the proof of Theorem 5.7.1 into two parts:

(I) We show that, if the maximal hypersurface M is complete with respect to the metric $\mathfrak{H}$ and if the norm of its Ricci curvature $\|Ric\|_{\mathfrak{H}}$ is bounded above then M must be an elliptic paraboloid.

(II) We use Hofer's Lemma (see section 5.5) to verify that $\|Ric\|_{\mathfrak{H}}$ must be bounded.

5.7.1 *Proof of Part I*

First we prove the following lemma:

Lemma 5.7.2. *Let $x : M \to \mathbb{R}^{n+1}$ be a locally strongly convex affine maximal hypersurface, which is given as graph of a locally strongly convex function:*

$$x^{n+1} = f(x^1, \cdots, x^n).$$

If M is complete with respect to the metric $\mathfrak{H}$, and if there is a constant $N > 0$ such that $\|Ric\|_{\mathfrak{H}} \leq N$ everywhere then, for dimension $n = 2$ or $n = 3$, M must be an elliptic paraboloid.

Proof. For the structure of the following proof recall section 4.5.5.
Let $p_0 \in M$ be an arbitrary fixed point. Adding a linear function and by an appropriate parameter transformation, we may assume that p_0 has coordinates $(0, \cdots, 0)$

and

$$f(0) = 0, \quad f_i(0) = 0, \quad f_{ij}(0) = \delta_{ij}.$$

Consider the function

$$F = (a^2 - r^2)^2 \Phi$$

defined on $\bar{B}_a(p_0, \mathfrak{H})$. Then, at p^*,

$$F_{,i} = 0, \qquad \sum F_{,ii} \leq 0,$$

where " , " denotes the covariant differentiation with respect to the Calabi metric. We calculate both expressions explicitly

$$\frac{\Phi_{,i}}{\Phi} - \frac{2(r^2)_{,i}}{a^2 - r^2} = 0, \tag{5.7.1}$$

$$\frac{\Delta \Phi}{\Phi} - \frac{\sum (\Phi_{,i})^2}{\Phi^2} - \frac{2\|\text{grad } r^2\|^2}{(a^2-r^2)^2} - \frac{2\Delta(r^2)}{a^2-r^2} \leq 0. \tag{5.7.2}$$

We insert (5.7.1) into (5.7.2) and get

$$\frac{\Delta \Phi}{\Phi} \leq \frac{6\|\text{grad } r^2\|^2}{(a^2-r^2)^2} + \frac{2\Delta(r^2)}{a^2-r^2} = \frac{24r^2}{(a^2-r^2)^2} + \frac{4}{a^2-r^2} + \frac{4r\Delta r}{a^2-r^2}. \tag{5.7.3}$$

Recall that $(M, \mathfrak{H})$ is a complete Riemann manifold with Ricci curvature bounded from below by a constant $-N$, $(N > 0)$. We apply the Laplacian Comparison Theorem and have

$$r\Delta r \leq (n-1)(1 + \sqrt{N} r).$$

Consequently, from (5.7.3), it follows that

$$\frac{\Delta \Phi}{\Phi} \leq \frac{24r^2}{(a^2-r^2)^2} + \frac{4n}{a^2-r^2} + \frac{4(n-1)\sqrt{N} \cdot r}{a^2-r^2}. \tag{5.7.4}$$

Case $n = 3$. In Proposition 4.5.2, choose $\beta = \frac{n-2}{2}$, $\delta = 0$, then we have

$$\Delta \Phi \geq \frac{3}{4} \frac{\|\text{grad } \Phi\|^2}{\Phi} - \frac{3}{2} \sum \Phi_{,i} \frac{\rho_{,i}}{\rho} + \frac{1}{6} \Phi^2. \tag{5.7.5}$$

Thus by (5.7.1) and the inequality of Schwarz we have

$$\begin{aligned} \frac{\Delta \Phi}{\Phi} &\geq \frac{3}{4} \sum \frac{(\Phi_{,i})^2}{\Phi^2} - \frac{3}{2} \sum \frac{\Phi_{,i}}{\Phi} \cdot \frac{\rho_{,i}}{\rho} + \frac{1}{6} \Phi \\ &\geq -6 \sum \frac{(\Phi_{,i})^2}{\Phi^2} + \frac{1}{12} \Phi \geq -\frac{100r^2}{(a^2-r^2)^2} + \frac{1}{12} \Phi. \end{aligned} \tag{5.7.6}$$

Insert (5.7.6) into (5.7.4); this gives

$$\Phi \leq \frac{1488r^2}{(a^2-r^2)^2} + \frac{144}{a^2-r^2} + \frac{96\sqrt{N} \cdot r}{a^2-r^2}. \tag{5.7.7}$$

Multiply both sides of (5.7.7) by $(a^2 - r^2)^2$. We obtain, at p^*,

$$(a^2 - r^2)^2 \Phi \leq b_1 a^2 + b_2 a^3, \tag{5.7.8}$$

for some positive constants b_1 and b_2.

Case $n = 2$. Choose $\beta = \frac{n-2}{2}$, $\delta = 0$ in Proposition 4.5.2, then we have

$$\Delta \Phi \geq \frac{\|\text{grad } \Phi\|^2}{\Phi} + \Phi^2. \tag{5.7.9}$$

Similar to the case $n = 3$, it is easy to verify that (5.7.8) holds also for $n = 2$. Hence, at any interior point of $\bar{B}_a(p_0, \mathfrak{H})$, we have

$$\Phi \le b_1 \frac{1}{a^2(1 - \frac{r^2}{a^2})^2} + b_2 \frac{1}{a(1 - \frac{r^2}{a^2})^2}.$$

Let $a \to \infty$, then

$$\Phi \equiv 0. \tag{5.7.10}$$

It follows that

$$\det\left(\frac{\partial^2 f}{\partial x^i \partial x^j}\right) = 1.$$

Thus the Calabi metric and the Blaschke metric satisfy

$$\mathfrak{H} = G.$$

This means that M is an affine complete parabolic affine hypersphere. By Theorem 4.6.1 we conclude that M must be an elliptic paraboloid. ■

5.7.2 *Proof of Part II: Affine blow-up analysis*

Now we want to show that there is a constant $N > 0$ such that $\|Ric\|_{\mathfrak{H}} \le N$ everywhere. To this end, we need Hofer's Lemma (see section 5.5).

Assume that $\|Ric\|_{\mathfrak{H}}$ is not bounded above. Then there is a sequence of points $p_\ell \in M$ such that $\|Ric\|_{\mathfrak{H}}(p_\ell) \to \infty$. Let $\bar{B}_1(p_\ell, \mathfrak{H})$ be the closed geodesic ball with center p_ℓ and radius 1. Consider a family $\Psi(\ell) : \bar{B}_2(p_\ell, \mathfrak{H}) \to \mathbb{R}$ of functions, $\ell \in \mathbb{N}$, defined by

$$\Psi(\ell) := \|Ric\|_{\mathfrak{H}} + \Phi + 4n(n-1)J, \tag{5.7.11}$$

where Φ and J are defined in section 3.3.4. In this subsection, if no confusion is possible, we simplify the notation again and write Ψ instead of $\Psi(\ell)$. We use Hofer's Lemma with $\Psi^{\frac{1}{2}}$ and find a sequence of points q_ℓ and positive numbers ϵ_ℓ such that

$$\Psi^{\frac{1}{2}}(x) \le 2\Psi^{\frac{1}{2}}(q_\ell), \quad \forall\ x \in \bar{B}_{\epsilon_\ell}(q_\ell, \mathfrak{H}), \tag{5.7.12}$$

$$\epsilon_\ell \Psi^{\frac{1}{2}}(q_\ell) \ge \tfrac{1}{2}\Psi^{\frac{1}{2}}(p_\ell) \to \infty. \tag{5.7.13}$$

The restriction of the hypersurface x to the balls $\bar{B}_{\epsilon_\ell}(q_\ell, \mathfrak{H})$ defines a family $M(\ell)$ of maximal hypersurfaces. For every ℓ, we normalize $M(\ell)$ as follows:

Step 1. By adding a linear function and by an appropriate coordinate transformation we may assume that q_ℓ has coordinates $(0, \cdots, 0)$ and

$$f(0) = 0, \ \ \text{grad}\, f(0) = 0.$$

We take a parameter transformation:

$$\widehat{x}^i(\ell) = \sum a^i_j(\ell) x^j(\ell), \tag{5.7.14}$$

where $a_i^j(\ell)$ are constants. Choosing $a_i^j(\ell)$ appropriately and using an obvious notation $\widehat{f}, \widehat{\Psi}$, we may assume that, for every ℓ, we have $\widehat{f}_{ij}(0) = \delta_{ij}$. Note that, under the parameter transformation (5.7.14), $\widehat{\Psi}$ is invariant.

Step 2. We take an affine transformation by

$$\widetilde{x}^i(\ell) = a(\ell)\widehat{x}^i(\ell), \quad 1 \le i \le n,$$
$$\widetilde{x}^{n+1}(\ell) = \lambda(\ell)\widehat{x}^{n+1}(\ell),$$

where $\lambda(\ell)$ and $a(\ell)$ are constants. It is easy to verify that each $\widetilde{M}(\ell)$ again is a locally strongly convex maximal hypersurface. Now we choose $\lambda(\ell) = a(\ell)^2 = \widehat{\Psi}(q_\ell)$. Using again an obvious notation $\widetilde{f}$, $\widetilde{\Psi}$, one can see that

$$\widetilde{f}_{ij}(\ell) = \widehat{f}_{ij}(\ell), \qquad \widetilde{\Psi}(\ell) = \tfrac{1}{\lambda(\ell)}\widehat{\Psi}(\ell). \tag{5.7.15}$$

The first equation in (5.7.15) is trivial. We calculate the second one. From the definition of Φ and Ric (see section 3.3.4) we easily get

$$\widetilde{\Phi} = \tfrac{1}{\lambda(\ell)}\widehat{\Phi}, \qquad \|\widetilde{Ric}\|_{\tilde{\mathfrak{H}}} = \tfrac{1}{\lambda(\ell)}\|\widehat{Ric}\|_{\hat{\mathfrak{H}}}, \qquad \widetilde{J} = \tfrac{1}{\lambda(\ell)}\widehat{J}.$$

Then the second equality in (5.7.15) follows.

We denote $\bar{B}_a(q_\ell, \tilde{\mathfrak{H}}) := \{x \in \widetilde{M}(\ell) \mid \widetilde{r}(\ell)(q_\ell, x) \le a\}$, where $\widetilde{r}(\ell)$ is the geodesic distance function with respect to the metric $\tilde{\mathfrak{H}}$ on $\widetilde{M}(\ell)$. Then $\widetilde{\Psi}(\ell)$ is defined on the geodesic ball $\bar{B}_{d(\ell)}(q_\ell, \tilde{\mathfrak{H}})$ with $d(\ell) = \epsilon_\ell \Psi^{\frac{1}{2}}(q_\ell) \ge \frac{1}{2}\Psi^{\frac{1}{2}}(p_\ell) \to \infty$. From (5.7.12) to (5.7.15) we have

$$\widetilde{\Psi}(q_\ell) = 1,$$

$$\widetilde{\Psi}(x) \le 4, \quad \forall x \in \bar{B}_{d(\ell)}(q_\ell, \tilde{\mathfrak{H}}). \tag{5.7.16}$$

We may identify the parametrization and write $(\xi_1, \cdots, \xi_n)$ for any index ℓ. Then $\widetilde{f}(\ell)$ is a sequence of functions defined in a domain $\Omega(\ell)$ with $0 \in \Omega(\ell)$. Thus we have a sequence $\widetilde{M}(\ell)$ of maximal hypersurfaces, given by $\widetilde{f}(\ell)$, and the following relations:

$$\widetilde{f}(\ell)(0) = 0, \quad \tfrac{\partial \widetilde{f}(\ell)}{\partial \xi_i}(0) = 0, \quad \tfrac{\partial^2 \widetilde{f}(\ell)}{\partial \xi_i \xi_j}(0) = \delta_{ij}, \tag{5.7.17}$$

$$\widetilde{\Psi}(\ell)(0) = 1, \tag{5.7.18}$$

$$\widetilde{\Psi}(\ell)(p) \le 4, \quad \forall p \in \bar{B}_{d(\ell)}(0, \tilde{\mathfrak{H}}), \tag{5.7.19}$$

$$d(\ell) \to \infty, \quad \text{as} \quad \ell \to \infty. \tag{5.7.20}$$

To continue with the proof of Part II of Theorem 5.7.1, we need the following lemma:

Lemma 5.7.3. *Let M be an affine maximal hypersurface defined in a neighborhood of $0 \in \mathbb{R}^n$. Suppose that, with the notations from above,*

(*i*) $$f_{ij}(0) = \delta_{ij},$$

(ii) $$\|Ric\|_{\mathfrak{H}} + \Phi + 4n(n-1)J \leq 4.$$

Denote $\bar{B}_{\frac{1}{2n}}(0) := \{(\xi_1, \cdots, \xi_n) | \sum(\xi_i)^2 \leq \frac{1}{4n^2}\}$. *Then there is a constant* $C_1 > 0$ *such that, for* $(\xi_1, \cdots, \xi_n) \in \bar{B}_{\frac{1}{2n}}(0)$, *the following estimates hold:*

(1) $$\sum f_{ii} \leq 4n,$$

(2) $$\tfrac{1}{C_1} \leq \det(f_{ij}) \leq C_1.$$

(3) *Define* d_o *by* $d_o^2 := \frac{1}{7n^2(4n)^{n-1}C_1}$ *then* $\bar{B}_{d_o}(0, \mathfrak{H}) \subset \{\sum(\xi_i)^2 < \frac{1}{7n^2}\} \subset \bar{B}_{\frac{1}{2n}}(0)$, *where* $B_{d_o}(0, \mathfrak{H})$ *is the geodesic ball with center* 0 *and radius* d_o *with respect to the metric* $\mathfrak{H}$.

Proof of Lemma 5.7.3.
(1) Consider an arbitrary curve $\Gamma = \{\xi_1 = a_1 s, \cdots, \ \xi_n = a_n s \mid \sum a_i^2 = 1, \ s \geq 0\}$. From the assumptions we have

$$\sum f^{il} f^{jm} f^{kr} f_{ijk} f_{lmr} \leq 4, \quad \sum f_{ii}(0) = n.$$

Since $\sum f^{il} f^{jm} f^{kr} f_{ijk} f_{lmr}$ is independent of the choice of the coordinates $\xi_1, \ldots, \xi_n$, for any point $\xi(s)$ we may assume that $f_{ij} = \lambda_i \delta_{ij}$. Then

$$\sum f^{il} f^{jm} f^{kn} f_{ijk} f_{lmn} = \sum \tfrac{1}{\lambda_i \lambda_j \lambda_k} f_{ijk}^2 \geq \tfrac{\sum f_{ijk}^2}{(\sum f_{ii})^3}.$$

It follows that

$$\tfrac{\sum f_{iik}^2}{(\sum f_{ii})^3} \leq \tfrac{1}{(\sum f_{ii})^3} \sum f_{ijk}^2 \leq 4,$$

and hence

$$\begin{aligned} \frac{1}{(\sum f_{ii}(\xi(s)))^{\frac{3}{2}}} \frac{d(\sum f_{ii}(\xi(s)))}{ds} &= \frac{1}{(\sum f_{ii}(\xi(s)))^{\frac{3}{2}}} \sum f_{iik}(\xi(s)) a_k \\ &\leq \sqrt{n} \left(\tfrac{\sum f_{iik}^2(\xi(s))}{(\sum f_{ii}(\xi(s)))^3} \right)^{\frac{1}{2}} \left(\sum a_k^2 \right)^{\frac{1}{2}} \\ &\leq 2\sqrt{n}. \end{aligned}$$

Solving this differential inequality with $\sum f_{ii}(0) = n$, we get

$$\tfrac{1}{\sqrt{n}} - s\sqrt{n} \leq \tfrac{1}{(\sum f_{ii}(\xi(s)))^{\frac{1}{2}}}.$$

From the assumptions we have $s \leq \frac{1}{2n}$, then **(1)** follows.

(2) Consider again an arbitrary curve

$$\Gamma = \{\xi_1 = a_1 s, \cdots, \xi_n = a_n s \mid \sum a_i^2 = 1, \ s \geq 0\}.$$

From the assumptions we have

$$\tfrac{\sum f^{ij} \rho_i \rho_j}{\rho^2} \leq 4.$$

It follows that

$$\tfrac{\sum (\rho_i)^2}{(\sum f_{ii}) \rho^2} \leq 4.$$

By **(1)** we get

$$\frac{1}{\rho}\frac{d\rho(\xi(s))}{ds} \leq 4\sqrt{n}.$$

Solving this differential inequality with $\rho(0)=1$, we obtain

$$-4\sqrt{n}s \leq \ln\ \rho(\xi(s)) \leq 4\sqrt{n}s.$$

Recall that $s \leq \frac{1}{2n}$, then **(2)** follows.

(3) Denote by $\lambda_{min}, \lambda_{max}$ the minimal and maximal eigenvalues of (f_{ij}), resp. Then, from **(1)** and **(2)**, we have $\lambda_{max} \leq 4n$ and

$$\frac{1}{C_1} \leq \det(f_{ij}) \leq \lambda_{min}\lambda_{max}^{n-1} \leq (4n)^{n-1}\lambda_{min}. \tag{5.7.21}$$

Hence, by **(1)** and (5.7.21), the geodesic distance function r satisfies

$$4n\sum(\xi_i)^2 \geq r^2 \geq \frac{1}{C_1(4n)^{n-1}}\sum(\xi_i)^2, \tag{5.7.22}$$

and **(3)** follows. This finishes the proof of Lemma 5.7.3. ■

To prove Part II of Theorem 5.7.1, we apply Lemma 5.7.3 to verify the following Claim:

Claim: *The sequence $\tilde{f}(\ell)$ locally uniformly converges in C^∞ to a smooth function $\tilde{f}$ that we consider as graph function of a locally strongly convex hypersurface; this hypersurface is complete with respect to the Calabi metric and satisfies the maximal hypersurface equation with bounded $\tilde{\Psi} := \lim \tilde{\Psi}(\ell)$.*

Once the claim is proved, by Lemma 5.7.2, $\tilde{f}$ must be a quadric. Hence $\tilde{\Psi} \equiv 0$. But

$$\tilde{\Psi}(0) = \lim_{\ell\to\infty} \tilde{\Psi}(\ell)(0) = 1.$$

We get a contradiction. Thus we show that there is a constant $N > 0$ such that $\|Ric\|_{\mathfrak{H}} \leq N$ everywhere.

Proof of the Claim: Since $d(\ell) \to \infty$, we have $\bar{B}_{\frac{1}{2n}}(0) \subset \Omega(\ell)$ for ℓ big enough. In fact, by (5.7.22), the geodesic distance from 0 to the boundary of $\bar{B}_{\frac{1}{2n}}(0)$, with respect to the metric $\tilde{\mathfrak{H}}$ on $\widetilde{M}(\ell)$, is less than $\frac{1}{\sqrt{n}}$. By Lemma 5.7.3 and bootstrapping, we get a C^k-estimate, independent of ℓ, for any k. It follows that there is a ball $B_{\sqrt{C_2}}(0) := \{\sum(\xi_i)^2 \leq C_2\}$ and a subsequence (still indexed by ℓ) such that $\widetilde{f}(\ell)$ converges to $\widetilde{f}$ on this ball, and correspondingly all derivatives, where $C_2 < \frac{1}{4n^2}$ is very close to $\frac{1}{4n^2}$. Thus, as limit, we get a maximal hypersurface $\widetilde{M}$, defined on this ball $B_{\sqrt{C_2}}(0)$, which contains a geodesic ball $\bar{B}_{d_o}(0, \tilde{\mathfrak{H}})$. In the following we will extend the hypersurface $\widetilde{M}$, and inductively show the statement

(*) $\tilde{f}(\ell)$ uniformly converges to $\tilde{f}$ in $B_{m\frac{d_o}{2}}(0, \tilde{\mathfrak{H}})$, $m \in \mathbb{Z}^+$. Moreover, for each $m < \infty$, $B_{m\frac{d_o}{2}}(0, \tilde{\mathfrak{H}})$ is bounded in $\mathbb{R}^n$.

We verify the statement (*) by induction on m. Assume (*) is true for m. Set

$$B(m-1) := \partial B_{(m-1)\frac{d_o}{2}}(0, \tilde{\mathfrak{H}})$$

and

$$\lambda_{\ell,\min}(m-1) := \inf_{\xi\in B(m-1)} \lambda_{\min}(D^2\tilde{f}(\ell)(\xi)),$$

$$\lambda_{\ell,\max}(m-1) := \sup_{\xi\in B(m-1)} \lambda_{\max}(D^2\tilde{f}(\ell)(\xi)),$$

where $D^2\tilde{f}(\ell)$ denotes the Hessian matrix of the function $\tilde{f}(\ell)$. Since $B(m-1)$ is compact, there exists $\epsilon_{m-1} > 0$ such that

$$\epsilon_{m-1} \leq \lambda_{\ell,\min}(m-1) \leq \lambda_{\ell,\max}(m-1) \leq \epsilon_{m-1}^{-1}.$$

In fact, this is true for $\lim \ell = \infty$, then by convergence, this is true for all ℓ.

Now we fix an arbitrary $\xi_o \in B_{m-1}$. Consider the convergence of $\tilde{f}(\ell)$ in the neighborhood of ξ_o. Note that the sequence $\{\tilde{f}(\ell)\}_\ell$ converges in $B_{\frac{d_o}{2}}(\xi_o, \tilde{\mathfrak{H}})$. We want to show that it converges in $B_{d_o}(\xi_o, \tilde{\mathfrak{H}})$. Again, analogously to step 1 above, we assume that $\xi_o = 0$, $\text{grad}\tilde{f}(\ell)(0) = 0$. We consider a parameter transformation:

$$\widehat{\xi}_i(\ell) = \sum a_i^j(\ell)\xi_j(\ell), \tag{5.7.23}$$

where $a_i^j(\ell)$ are constants. Choosing $a_i^j(\ell)$ appropriately and using an obvious notation $\widehat{f}, \widehat{\Psi}$, we may assume that, for every ℓ, we have $\widehat{f}_{ij}(0) = \delta_{ij}$. Note that, under the parameter transformation (5.7.23), $\widehat{\Psi}$ is invariant. It is easy to verify that each $\widehat{M}(\ell)$ again is a locally strongly convex maximal hypersurface, and

$$\frac{\partial^2 \widehat{f}(\ell)}{\partial\widehat{\xi}_i\widehat{\xi}_j}(0) = \delta_{ij},$$
$$\widehat{\Psi}(\ell)(p) \leq 4, \quad \forall p \in \bar{B}_{d(\ell)}(0, \widehat{\mathfrak{H}}),$$
$$d(\ell) \to \infty, \quad \text{as} \quad \ell \to \infty.$$

Now we apply Lemma 5.7.3 again to conclude that $\widehat{f}(\ell)$ converges to $\widehat{f}$ in $B_{\sqrt{C_2}}(0)$ and hence on the geodesic ball $B_{d_o}(\xi_o, \widehat{\mathfrak{H}})$. Note that the upper and lower bounds of eigenvalues of $(a_i^j(\ell))$ only depend on $\lambda_{\ell,\min}(m-1)$ and $\lambda_{\ell,\max}(m-1)$, and hence on ϵ_{m-1}. Put $A_\infty := \lim_{\ell\to\infty}(a_i^j(\ell))$. Hence $A_\infty^{-1}(B_{\sqrt{C_2}}(0))$ is bounded.

Now, we reverse the above affine rescaling and conclude that:

$\tilde{f}(\ell)$ converges to $\tilde{f}$ in a bounded domain $D(\xi_o) := A_\infty^{-1}(B_{\sqrt{C_2}}(0))$ and thus on the subset $B_{d_o}(\xi_o, \tilde{\mathfrak{H}})$.

The choice of ξ_o was arbitrary. If $\xi = \xi_o$ runs over $B(m-1)$, we conclude that $\tilde{f}(\ell)$ uniformly converges to $\tilde{f}$ in $B_{(m+1)\frac{d_o}{2}}(0, \tilde{\mathfrak{H}})$. Moreover, it is easy to see that

$$B_{(m+1)\frac{d_o}{2}}(0, \tilde{\mathfrak{H}}) \subset B_{m\frac{d_o}{2}}(0, \tilde{\mathfrak{H}}) \bigcup_{\xi\in B(m-1)} D(\xi)$$

which is bounded; here $D(\xi)$ is defined in analogy to $D(\xi_o)$. This proves (*). Now from (*) the claim follows immediately. This completes the proof of Part II and thus of Theorem 5.7.1. ∎

We pose the following problem for higher dimension:

Problem 5.7.4. *Let $x : M \to \mathbb{R}^{n+1}$ be a locally strongly convex hypersurface, given as graph of a convex function*

$$x^{n+1} = f(x^1, \cdots, x^n),$$

defined on a domain $\Omega \subset \mathbb{R}^n$. If $x(M)$ is an affine maximal hypersurface and if $x(M)$ is complete with respect to the metric $\mathfrak{H}$, is it an elliptic paraboloid ?

5.8 Another Method of Proof for some Fourth Order PDEs

In this section we shall prove Bernstein properties for complete hypersurfaces of dimension $n \geq 2$, satisfying the PDE (4.5.9), where we consider two different completeness conditions, namely Calabi completeness in Theorem 5.8.1 and Euclidean completeness in Theorem 5.8.2.
For the proof of Theorem 5.7.1 we used analytic blow up techniques. For the proof of the following Theorem 5.8.1 we follow ideas of A.M. Li and F. Jia, introduced in [45]; there both authors studied the constant affine mean curvature equation. In our proof below we give a lower bound for the Ricci curvature, calculating with the Calabi metric. Then we apply the Laplacian Comparison Theorem.

Theorem 5.8.1. *Let $f(x^1, ..., x^n)$ be a strictly convex C^∞-function defined on a convex domain $\Omega \in \mathbb{R}^n$ satisfying the PDE (4.5.9). Define*

$$M := \{(x, f(x)) \mid x^{n+1} = f(x), x := (x^1, ..., x^n) \in \Omega\}.$$

If M is complete with respect to the Calabi metric $\mathfrak{H}$ and $\beta \notin \left[-\frac{(n+2)(n-1)}{4\sqrt{n}} - 1, \frac{(n+2)(n-1)}{4\sqrt{n}} - 1\right]$ then M must be an elliptic paraboloid.

Proof. Recall the beginning of the proof of Lemma 5.7.2. Define the geodesic distance from p_o with respect to the Calabi metric: $a^* = r(p_0, p^*)$. We discuss the two cases $p^* \neq p_0$ and $p^* = p_0$.
1. In case $p^* \neq p_0$ we have $a^* > 0$. Let $\bar{B}_{a^*}(p_0, \mathfrak{H}) := \{p \in M \mid r(p_0, p) \leq a^*\}$. We choose $\delta = 0$ and

$$\beta \notin \left[-\frac{(n+2)(n-1)}{4\sqrt{n}} - 1, \frac{(n+2)(n-1)}{4\sqrt{n}} - 1\right]$$

in Proposition 4.5.2, and apply the maximum principle, then we have

$$\max_{\bar{B}_{a^*}(p_0,\mathfrak{H})} \Phi = \max_{\partial \bar{B}_{a^*}(p_0,\mathfrak{H})} \Phi.$$

Observe that $a^2 - r^2 = a^2 - a^{*2}$ on $\partial \bar{B}_{a^*}(p_0, \mathfrak{H})$, thus it follows that

$$\max_{\bar{B}_{a^*}(p_0,\mathfrak{H})} \Phi = \Phi(p^*).$$

Consider $p \in \bar{B}_{a^*}(p_0, \mathfrak{H})$; we choose an affine coordinate neighborhood $\{U, \varphi\}$ with $p \in U$ such that

$$R_{ij}(p) = 0, \quad \text{for} \quad i \neq j, \quad \text{and} \quad f_{ij}(\varphi(p)) = \frac{\partial^2 f}{\partial x_i \partial x_j}(\varphi(p)) = \delta_{ij}, \quad 1 \leq i, j \leq n,$$

in U. From (4.6.3) we get

$$R_{ii}(p) \geq -\tfrac{(n+2)^2}{16}\,\Phi(p) \geq -\tfrac{(n+2)^2}{16}\,\Phi(p^*).$$

We apply the Laplacian Comparison Theorem from section 1.3 and obtain

$$r\Delta r \leq (n-1)\left(1 + \tfrac{n+2}{4}\sqrt{\Phi(p^*)}\cdot r\right). \tag{5.8.1}$$

2. In case $p^* = p_0$ we have $r(p_0, p^*) = 0$. Consequently, from (5.7.3), (5.8.1) and the inequality of Schwarz it follows that

$$\tfrac{\Delta\Phi}{\Phi} \leq C_1\,\tfrac{a^2}{(a^2-r^2)^2} + \epsilon\Phi, \tag{5.8.2}$$

where $\epsilon > 0$ is a small constant to be determined later, and C_1 is a positive constant depending only on n and ϵ. (5.8.2) gives an upper estimate for the expression $\frac{\Delta\Phi}{\Phi}$. In the next step we calculate a lower estimate. Namely, by Proposition 4.5.2 with $\delta = 0$ and the inequality of Schwarz we have

$$\tfrac{\Delta\Phi}{\Phi} \geq -16\tfrac{(\beta+1)^2(n-2)^2}{(n-1)^2\epsilon}\,\tfrac{a^2}{(a^2-r^2)^2} + \left(\tfrac{2(\beta+1)^2}{n-1} - \tfrac{(n+2)^2(n-1)}{8n} - \epsilon\right)\Phi; \tag{5.8.3}$$

here we used (5.7.1). We combine (5.8.2) with (5.8.3) and have

$$F(\beta,\epsilon)\Phi \leq C_2\tfrac{a^2}{(a^2-r^2)^2},$$

where

$$F(\beta,\epsilon) := \tfrac{2(\beta+1)^2}{n-1} - \tfrac{(n+2)^2(n-1)}{8n} - 2\epsilon$$

and C_2 is a positive constant depending only on n and ϵ. We may choose a sufficiently small number $\epsilon(\beta) > 0$ s. t. $F(\beta,\epsilon) > 0$. Hence, at p^*,

$$\Phi \leq C_3\tfrac{a^2}{(a^2-r^2)^2},$$

for some positive number C_3 depending only on n and β. Thus, at any interior point of $\bar{B}_a(p_0, \mathfrak{H})$, we obtain

$$\Phi \leq C_3\tfrac{a^2}{(a^2-r^2)^2}.$$

For $a \to \infty$ we get

$$\Phi \equiv 0.$$

This means that M is an affine complete parabolic hypersphere. We apply Theorem 4.6.1 and conclude that M must be an elliptic paraboloid. This completes the proof of Theorem 5.8.1. ■

Remark. For affine maximal hypersurfaces we know that $\beta = \frac{n-2}{2}$ in the PDE (4.5.9). It is easy to check that Theorem 5.7.1 is a special case of Theorem 5.8.1. Following Li and Jia's idea, several authors obtained similar results, see [44], [71] and [101] .

In [46] the authors proved the following Bernstein property. However in [46] the calculations are very complicated. Later, in [102], we gave a relatively simple proof, using the Calabi metric.

Theorem 5.8.2. *Let $x : M \to \mathbb{R}^{n+1}$ be a locally strongly convex hypersurface, which is given as graph of a locally strongly convex function:*

$$x^{n+1} = f(x^1, \cdots, x^n),$$

defined on $\mathbb{R}^n$. If f satisfies the PDE (4.5.9) then there is a positive constant $K(n)$ depending only on the dimension n, such that, if $|\beta| \geq K(n)$ then f must be a quadratic polynomial.

To structure the proof of Theorem 5.8.2, we first prove the Propositions 5.8.3 and 5.8.4 below.

For any fixed convex domain $\Omega \subset \mathbb{R}^n$, and any constant $C > 0$, let $f \in \mathcal{S}(\Omega, C)$ (see section 1.1.4). Assume that the function Φ does not vanish identically on Ω. For the sake of simplicity, we introduce the following notations:

$$\mathcal{A} := \max_{\Omega} \left\{ \exp\left\{ -\tfrac{m}{C-f} \right\} \Phi \right\},$$

$$K := \tfrac{1}{|\beta| \mathcal{A}} \exp\left\{ -\tfrac{m}{C-f} \right\} \Phi,$$

$$\mathcal{D} := \max_{\Omega} \left\{ \exp\left\{ -\tfrac{m}{C-f} + K \right\} g^2 \|\text{grad} f\|^2 \right\},$$

where the function g is defined in Lemma 5.6.3, and m is a positive constant to be determined later. We may assume that $|\beta| \geq 1$.

Proposition 5.8.3. *On a convex domain Ω, assume that $f \in \mathcal{S}(\Omega, C)$ satisfies the PDE (4.5.9). Then there is a positive constant $K_1(n)$, depending only on n, such that, if $|\beta| > K_1(n)$ then the following estimate holds:*

$$\mathcal{A} \leq a_0 (\mathcal{D} + \tfrac{1}{C}),$$

where

$$a_0 := \tfrac{16n^2}{(\beta+1)^2 - (n+2)^3} > 0.$$

Proof. To prove this proposition we consider the function

$$F := \exp\left\{ -\tfrac{m}{C-f} \right\} \Phi,$$

defined on Ω, where m is a positive constant to be determined later. Clearly, F attains its supremum at some interior point p^*. Around p^* choose a local orthonormal frame field with respect to the Calabi metric $\mathfrak{H}$. Recall the definitions of g and g' from Lemma 5.6.3. Then, at p^*,

$$\tfrac{\Phi_{,i}}{\Phi} - g f_{,i} = 0, \tag{5.8.4}$$

$$\tfrac{\Delta \Phi}{\Phi} - \tfrac{\|\text{grad } \Phi\|^2}{\Phi^2} - g' \|\text{grad} f\|^2 - g \Delta f \leq 0. \tag{5.8.5}$$

Proposition 4.5.2 with $\delta = 0$, formula (5.8.4) and the inequality of Schwarz give:

$$\begin{aligned} \tfrac{\Delta \Phi}{\Phi} - \tfrac{\|\text{grad } \Phi\|^2}{\Phi^2} &\geq -\tfrac{1}{2} \tfrac{\|\text{grad } \Phi\|^2}{\Phi^2} - 2\tfrac{(\beta+1)(n-2)}{n-1} \sum \tfrac{\Phi_{,i}}{\Phi} \tfrac{\rho_{,i}}{\rho} + \left(\tfrac{2(\beta+1)^2}{n} - \tfrac{(n+2)^2}{8} \right) \Phi \\ &\geq -2n \tfrac{\|\text{grad } \Phi\|^2}{\Phi^2} + \left(\tfrac{(\beta+1)^2}{n} - \tfrac{(n+2)^2}{8} \right) \Phi \\ &\geq -2n g^2 \|\text{grad} f\|^2 + \left(\tfrac{(\beta+1)^2}{n} - \tfrac{(n+2)^2}{8} \right) \Phi. \end{aligned} \tag{5.8.6}$$

Again we apply the inequality of Schwarz; it follows that

$$g\Delta f \leq ng + \tfrac{(n+2)^2}{8n}\Phi + \tfrac{n}{2}g^2\|\mathrm{grad} f\|^2. \tag{5.8.7}$$

Choose $m \geq 4C$, then $g' \leq g^2$. Insert (5.8.6) and (5.8.7) into (5.8.5); we obtain

$$\left(\tfrac{(\beta+1)^2}{n} - \tfrac{(n+2)^2}{4}\right)\Phi - 4ng^2\|\mathrm{grad} f\|^2 - ng \leq 0. \tag{5.8.8}$$

Put

$$a_0 := \tfrac{16n^2}{(\beta+1)^2-(n+2)^3},$$

then there clearly exists a constant $K_1(n)$ such that $a_0 > 0$ in case that $|\beta| > K_1(n)$. Therefore

$$\Phi \leq a_0 g^2\|\mathrm{grad} f\|^2 + a_0 g. \tag{5.8.9}$$

Multiply both sides of (5.8.9) with the factor $\exp\left\{-\frac{m}{C-f} + K\right\}(p^*)$, and use $K(p^*) = \frac{1}{|\beta|}$; this gives the asserted inequality

$$\mathcal{A} \leq a_0(\mathcal{D} + \tfrac{1}{C}).$$

Proposition 5.8.3 is proved. ∎

Proposition 5.8.4 *On a convex domain Ω, assume that $f \in \mathcal{S}(\Omega, C)$ satisfies the PDE (4.5.9). Then there is a positive constant $K_2(n)$, depending only on n, such that, if $|\beta| > K_2(n)$ then the following estimates hold:*

$$\mathcal{A} \leq \tfrac{d_1}{C}, \quad \mathcal{D} \leq \tfrac{d_1}{C}$$

for some constant $d_1 > 0$ depending only on β and n.

Proof. First we use Lemma 5.6.3 and (5.6.10) to obtain the following estimates:

$$\begin{aligned}&\tfrac{4-6n\delta}{n-1}\sum(f_{,1i})^2 + (n+2)\tfrac{\rho_{,11}}{\rho}(f_{,1})^2 - \tfrac{3(n+2)^3}{8\delta}\Phi(f_{,1})^2 - 4n - \tfrac{4}{\delta(n-1)^2}(\Delta f)^2\\ &\quad + \left(\Delta\tau^\sharp - g'(f_{,1})^2 - g\Delta f + \tfrac{\Delta Q}{Q} - \tfrac{\sum(Q_{,i})^2}{Q^2}\right)(f_{,1})^2 \leq 0\end{aligned} \tag{5.8.10}$$

for any number $\delta \in (0, \frac{1}{2})$. Put $\tau^\sharp := K$, $Q := g^2$ in (5.8.10). Assume that F, defined by (5.6.6), attains its supremum at some point q^*. Then, from $F_{,i}(q^*) = 0$, we have at q^*:

$$\left(-gf_{,i} + \tfrac{4}{C-f}f_{,i} + K_{,i}\right)(f_{,1})^2 + 2\sum f_{,j}f_{,ji} = 0. \tag{5.8.11}$$

We use formula (5.8.10) with $\delta = \frac{1}{8n}$; then we know that, at q^*,

$$\begin{aligned}&\tfrac{3}{n-1}\sum(f_{,1i})^2 + (n+2)\tfrac{\rho_{,11}}{\rho}(f_{,1})^2 - 10(n+2)^4\Phi(f_{,1})^2 - 100n^3 + \Delta K(f_{,1})^2\\ &\quad - \left(g'(f_{,1})^2 + ng + \tfrac{n+2}{2}\left(g - \tfrac{4}{C-f}\right)\tfrac{\rho_{,1}}{\rho}f_{,1}\right)(f_{,1})^2 \leq 0.\end{aligned} \tag{5.8.12}$$

In the following we compute the two terms $\sum(f_{,1i})^2$ and ΔK, respectively. For this we note that $K \leq \frac{1}{|\beta|}$, and also the elementary inequality

$$(a+b)^2 \geq (1-\varepsilon)a^2 - (\tfrac{1}{\varepsilon} - 1)b^2, \quad \text{for any } \varepsilon > 0. \tag{5.8.13}$$

In (5.8.13) choose

$$a = \left(g - \tfrac{4}{C-f}\right) f_{,1}, \quad b = K_{,1}, \quad \varepsilon = \tfrac{5K}{1+5K}.$$

Then, from (5.8.11), it follows that

$$\begin{aligned} 4\sum (f_{,1i})^2 &= \sum \left(g f_{,i} - \tfrac{4}{C-f} f_{,i} - K_{,i}\right)^2 (f_{,1})^2 \\ &\geq \tfrac{|\beta|}{|\beta|+5} \left(g - \tfrac{4}{C-f}\right)^2 (f_{,1})^2 \sum (f_{,i})^2 - \tfrac{1}{5} \tfrac{\sum (K_{,i})^2}{K} (f_{,1})^2. \end{aligned} \tag{5.8.14}$$

Next we find that

$$\begin{aligned} K_{,i} =& K\left(\tfrac{\Phi_{,i}}{\Phi} - g f_{,i}\right), \\ \Delta K =& \tfrac{\sum (K_{,i})^2}{K} + K\left(\tfrac{\Delta \Phi}{\Phi} - \tfrac{\|\text{grad } \Phi\|^2}{\Phi^2}\right) - K\left(g'(f_{,1})^2 + g\Delta f\right). \end{aligned}$$

By virtue of Proposition 4.5.2 with $\delta = \frac{1}{8n}$, (5.8.7) and the inequality of Schwarz we get

$$\begin{aligned} \Delta K \geq& \tfrac{\sum (K_{,i})^2}{K} + \tfrac{1}{4n} \tfrac{K}{\Phi} \sum \tfrac{(\rho_{,ij})^2}{\rho^2} - \tfrac{1}{|\beta|}\left(\left(\tfrac{n}{2} g^2 + g'\right)(f_{,1})^2 + ng\right) \\ &+ K\left(\tfrac{15-8n}{16(n-1)} \tfrac{\|\text{grad } \Phi\|^2}{\Phi^2} - \tfrac{8n(n-2)(\beta+1)+(\beta+n)}{4n(n-1)} \sum \tfrac{\Phi_{,i}}{\Phi} \tfrac{\rho_{,i}}{\rho} - (n+2)^2 \Phi\right) \\ \geq& \tfrac{\sum (K_{,i})^2}{K} + \tfrac{1}{4n} \tfrac{K}{\Phi} \sum \tfrac{(\rho_{,ij})^2}{\rho^2} - \tfrac{1}{|\beta|}\left(\left(\tfrac{n}{2} g^2 + g'\right)(f_{,1})^2 + ng\right) \\ &- K\left(\tfrac{1}{2} \tfrac{\|\text{grad } \Phi\|^2}{\Phi^2} + \left(n(\beta+1)^2 + (n+2)^2\right)\Phi\right). \end{aligned} \tag{5.8.15}$$

Since

$$\tfrac{6}{10} \tfrac{\sum (K_{,i})^2}{K} = \tfrac{6}{10} K\left(\tfrac{\Phi_{,i}}{\Phi} - g f_{,i}\right)^2 \geq \tfrac{1}{2} K \tfrac{\|\text{grad } \Phi\|^2}{\Phi^2} - 3Kg^2(f_{,1})^2, \tag{5.8.16}$$

we have

$$\begin{aligned} \Delta K \geq& \tfrac{1}{4n} \tfrac{K}{\Phi} \sum \tfrac{(\rho_{,ij})^2}{\rho^2} + \tfrac{2}{5} \tfrac{\sum (K_{,i})^2}{K} - a_1 \tfrac{1}{|\beta|} \Phi \\ &- \tfrac{1}{|\beta|}\left(\left((n+3)g^2 + g'\right)(f_{,1})^2 + ng\right), \end{aligned} \tag{5.8.17}$$

where

$$a_1 := n(\beta+1)^2 + (n+2)^2.$$

The inequality of Schwarz gives

$$(n+2) \tfrac{\rho_{,11}}{\rho} (f_{,1})^2 \leq \tfrac{1}{4n} \tfrac{K}{\Phi} \sum \tfrac{(\rho_{,ij})^2}{\rho^2} (f_{,1})^2 + n(n+2)^2 \tfrac{\Phi}{K} (f_{,1})^2. \tag{5.8.18}$$

Insert (5.8.14), (5.8.17) and (5.8.18) into (5.8.12) and use the inequality of Schwarz again; we get

$$\begin{aligned} &\tfrac{|\beta|}{2(n-1)(|\beta|+5)} \left(g - \tfrac{4}{C-f}\right)^2 (f_{,1})^4 - a_2 \Phi (f_{,1})^2 - n(n+2)^2 \tfrac{\Phi}{K} (f_{,1})^2 \\ &\quad - \left(\tfrac{n+3}{|\beta|} g^2 + \tfrac{1+|\beta|}{|\beta|} g'\right)(f_{,1})^4 - 2ng(f_{,1})^2 - 100n^3 \leq 0, \end{aligned} \tag{5.8.19}$$

where

$$a_2 := \tfrac{a_1}{|\beta|} + 11(n+2)^4.$$

Choose $m \geq 2000(n-1)C$ so that

$$g' < \tfrac{|\beta|^2}{16(n-1)(|\beta|+5)^2} g^2, \qquad \tfrac{4}{C-f} < \tfrac{1}{2} g.$$

Thus

$$\begin{aligned} &\left(\tfrac{|\beta|}{16(n-1)(|\beta|+5)} - \tfrac{n+3}{|\beta|}\right) g^2 (f_{,1})^4 - a_2 \Phi (f_{,1})^2 \\ &\qquad - n(n+2)^2 \tfrac{\Phi}{K} (f_{,1})^2 - 2ng(f_{,1})^2 - 100n^3 \leq 0. \end{aligned} \tag{5.8.20}$$

Choose $|\beta|$ large enough such that

$$a_3 := \tfrac{16(n-1)(|\beta|+5)|\beta|}{|\beta|^2 - 16(n+3)^2(|\beta|+5)} > 0.$$

Then

$$\tfrac{1}{2} g^2 (f_{,1})^4 - a_2 a_3 \Phi (f_{,1})^2 - n(n+2)^2 a_3 \tfrac{\Phi}{K} (f_{,1})^2 - 2n^2 a_3^2 - 100n^3 a_3 \leq 0. \tag{5.8.21}$$

Multiply both sides of (5.8.21) with the factor $\exp\left\{-\frac{2m}{C-f} + 2K\right\} g^2(q^*)$; we have

$$\begin{aligned} \mathcal{D}^2 \leq & a_3 \exp\left\{-\tfrac{2m}{C-f} + 2K\right\} g^2 \left\{2a_2 \Phi (f_{,1})^2 + 2n(n+2)^2 \tfrac{\Phi}{K} (f_{,1})^2\right\} \\ & + \exp\left\{-\tfrac{2m}{C-f} + 2K\right\} g^2 \left\{4n^2 a_3^2 + 200n^3 a_3\right\} \\ \leq & a_3 \exp\left\{\tfrac{1}{|\beta|}\right\} \left(2a_2 + 2n(n+2)^2 |\beta|\right) \mathcal{AD} + a_4 \\ \leq & a_5 \mathcal{AD} + a_4, \end{aligned} \tag{5.8.22}$$

where

$$a_4 := \tfrac{160}{m^2}(4n^2 a_3^2 + 200n^3 a_3), \qquad a_5 := 6a_3(a_2 + n(n+2)^2 |\beta|).$$

Recall the definition of a_0 in Proposition 5.8.3, and note that, when $|\beta|$ is sufficiently large,

$$2a_0 a_5 = \tfrac{192n^2}{(\beta+1)^2 - (n+2)^3} \left(a_2 + |\beta| n(n+2)^2\right) a_3 \sim \tfrac{1}{|\beta|}. \tag{5.8.23}$$

Thus there exists a positive constant $K_2(n)$, depending only on n, such that, if

$$|\beta| > K_2(n), \quad a_3 > 0, \quad 2a_0 a_5 < \tfrac{1}{2},$$

we get

$$\mathcal{D} \leq \tfrac{a_6}{C},$$

where a_6 is a positive constant depending only on β and n. This together with Proposition 5.8.3 gives Proposition 5.8.4. ∎

Proof of Theorem 5.8.2. Let f be a strictly convex function satisfying (4.5.9). Modulo an affine transformation of $\mathbb{R}^{n+1}$, we can assume that $f(0) = 0$ and $f > 0$ on $\mathbb{R}^n \backslash \{0\}$. Consider the *sections* of f:

$$S_f(0, C) := \{p \in \mathbb{R}^n \mid f(p) < C\}, \quad \forall C > 0.$$

They are convex open domains in $\mathbb{R}^n$. For any point $p \in M$, choose a sufficiently large constant $C_0 > 0$ such that $p \in S_f(0, C_0)$. Then, for all $C \geq C_0$, we have $p \in S_f(0, C_0) \subset S_f(0, C)$. From Proposition 5.8.4, we know that the inequality

$$\exp\left\{-\tfrac{m}{C-f}\right\}\Phi \leq \tfrac{d_1}{C}$$

holds on each section $S_f(0, C)$ with $C \geq C_0$. In particular, we have

$$\exp\left\{-\tfrac{m}{C-f(p)}\right\}\Phi(p) \leq \tfrac{d_1}{C}. \tag{5.8.24}$$

Now take $C \to +\infty$ in (5.8.24); it follows that $0 \leq \Phi(p) \leq 0$, which implies the equation $\Phi(p) = 0$. But the choice of $p \in M$ was arbitrary, thus Φ vanishes identically on M. This means that ρ =const and thus M is a Euclidean complete, parabolic affine hypersphere. An application of Pogorelov's theorem (see section 4.4) implies that M must be an elliptic paraboloid; this completes the proof of Theorem 5.8.2. ■

5.9 Euclidean Completeness and Calabi Completeness

Recall different notions of completeness given in section 4.2. In this section, under additional assumptions, we shall prove that, for a graph hypersurface $M = \{(x, f(x))\}$ the Euclidean completeness of M implies the Calabi completeness.

Remarks and Examples. (i) Generally, the notions of *Calabi completeness* and *Euclidean completeness* on M are *not* equivalent. For example, the global graph over $\mathbb{R}^n$ in $\mathbb{R}^{n+1}$, given by

$$h(x) = \exp\{x^1\} + \sum_{i=2}^{n}(x^i)^2$$

is Euclidean complete, but not Calabi complete.

(ii) Generally, the notions of *Calabi completeness* and *affine completeness* on M are *not* equivalent. For example, the one-sheeted hyperboloid (see [45])

$$f(x^1, x^2) = (1 + (x^1)^2 + (x^2)^2)^{\frac{1}{2}}$$

$$M := \{(x^1, x^2, f(x^1, x^2)) \mid (x^1, x^2) \in \mathbb{R}^2\}$$

is Euclidean complete and affine complete, but not Calabi complete.

Theorem 5.9.1. *Let $x^{n+1} = f(x)$ be a strictly convex function defined on a convex domain $\Omega \subset \mathbb{R}^n$ satisfying the PDE (5.6.3) with $L^\sharp =$ const. Assume that $M = \{(x, f(x)) \mid x \in \Omega\}$ is a Euclidean complete hypersurface and that Φ is bounded then M is complete with respect to the Calabi metric $\mathfrak{H}$.*

In [45] the authors prove such a conclusion for a locally strongly convex hypersurface with constant affine mean curvature, which is a special case of Theorem 5.9.1.

Namely Theorem 5.9.1 is more general than Theorem 1 of [45]. In the following we use the Calabi metric to prove Theorem 5.9.1.

Proof. Let $p \in M$ be any fixed point. Up to an affine transformation of $\mathbb{R}^{n+1}$, we may assume that p has coordinates $(0, \cdots, 0, 0)$ and

$$f(0) = 0, \quad \tfrac{\partial f}{\partial x^i}(0) = 0, \quad 1 \le i \le n.$$

The key point of the proof of Theorem 5.9.1 is to estimate $\frac{\|\text{grad } f\|^2}{(1+f)^2}$. We shall show that it is bounded if Φ is bounded. To estimate $\frac{\|\text{grad } f\|^2}{(1+f)^2}$, we consider the following function

$$F := \exp\left\{\tfrac{-m}{C-f} + \Psi\right\} \tfrac{\|\text{grad } f\|^2}{(1+f)^2} \tag{5.9.1}$$

defined on the *section* $S_f(0, C)$, where

$$\Psi := \exp\{\Phi\},$$

and m is a positive constant to be determined later. Clearly, F attains its supremum at some interior point p^* of $S_f(0, C)$. We can assume that $\|\text{grad } f\| > 0$ at p^*. Choose a local orthonormal frame field of the Calabi metric $e_1, \cdots, e_n$ on M such that, at p^*, $f_{,1} = \|\text{grad } f\| > 0$, $f_{,i} = 0$ $(i \ge 2)$. Then, at p^*,

$$F_{,i} = 0, \tag{5.9.2}$$

$$\sum F_{,ii} \le 0. \tag{5.9.3}$$

Now we calculate both expressions (5.9.2) and (5.9.3) explicitly. By (5.9.2) and (5.9.3), we have

$$2\sum f_{,j} f_{,ji} + \left(-g f_{,i} - 2\tfrac{f_{,i}}{1+f} + \Psi_{,i}\right) \sum (f_{,j})^2 = 0, \tag{5.9.4}$$

$$\begin{aligned} &2\sum (f_{,ij})^2 + 2\sum f_{,j} f_{,jii} + 2\sum \left(-g f_{,i} - 2\tfrac{f_{,i}}{1+f} + \Psi_{,i}\right) f_{,j} f_{,ji} \\ &\quad + \left[-g' \sum (f_{,i})^2 - g\Delta f + 2\tfrac{\sum (f_{,i})^2}{(1+f)^2} - 2\tfrac{\Delta f}{1+f} + \Delta\Psi\right] \sum (f_{,j})^2 \le 0, \end{aligned} \tag{5.9.5}$$

where, as before, g and g' are defined in Lemma 5.6.3.
Let us simplify (5.9.5). From (5.9.4) we have

$$2 f_{,1i} = \left(g f_{,i} + 2\tfrac{f_{,i}}{1+f} - \Psi_{,i}\right) f_{,1}. \tag{5.9.6}$$

Similar to (4.5.43), applying the inequality of Schwarz, we get

$$2\sum (f_{,ij})^2 \ge 2\left(\tfrac{n}{n-1} - \delta\right)(f_{,11})^2 + 4\sum_{i>1} (f_{,1i})^2 - \tfrac{2}{\delta(n-1)^2}(\Delta f)^2, \tag{5.9.7}$$

for any $0 < \delta < 1$. We insert (5.9.6) and (5.9.7) into (5.9.5) and obtain

$$\begin{aligned} &\left[2\left(\tfrac{n}{n-1} - \delta\right) - 4\right](f_{,11})^2 + 2\sum f_{,j} f_{,jii} - \tfrac{2}{\delta(n-1)^2}(\Delta f)^2 \\ &\quad + \left[-g'(f_{,1})^2 - g\Delta f + 2\tfrac{(f_{,1})^2}{(1+f)^2} - 2\tfrac{\Delta f}{1+f} + \Delta\Psi\right](f_{,1})^2 \le 0. \end{aligned} \tag{5.9.8}$$

Now we calculate $\Delta\Psi$. We use a calculation similar to Proposition 4.5.2 and obtain

$$\Delta\Phi \geq \tfrac{2\delta}{\rho^2}\sum(\rho_{,ij})^2 - C(n,\delta,\beta)\Phi^2, \tag{5.9.9}$$

where $C(n,\delta,\beta)$ is a positive constant depending only on n, δ and β. Then we get

$$\Psi_{,i} = \Psi\Phi_{,i}, \tag{5.9.10}$$

and

$$\Delta\Psi \geq \Psi\sum(\Phi_{,i})^2 + \Psi\tfrac{2\delta}{\rho^2}\sum(\rho_{,ij})^2 - C(n,\delta,\beta)\Psi\Phi^2. \tag{5.9.11}$$

Let us now compute the terms $\sum f_{,j}f_{,jii}$ from (5.9.8). An application of the Ricci identity shows that

$$2\sum f_{,j}f_{,jii} = 2\sum f_{,j}(\Delta f)_{,j} + 2\sum R_{ij}f_{,i}f_{,j}.$$

We use (3.3.4) and (4.5.5) to obtain

$$\begin{aligned}
2\sum f_{,j}f_{,jii} =& (n+2)\left(f_{,ij}\tfrac{\rho_{,i}}{\rho} + f_{,i}\tfrac{\rho_{,ij}}{\rho} - f_{,i}\tfrac{\rho_{,i}\rho_{,j}}{\rho^2}\right)f_{,j} + 2\sum R_{11}(f_{,1})^2 \\
=& (n+2)\left(\tfrac{\rho_{,11}}{\rho}(f_{,1})^2 - \tfrac{(\rho_{,1})^2}{\rho^2}(f_{,1})^2 + f_{,1i}f_{,1}\tfrac{\rho_{,i}}{\rho} - A_{11k}\tfrac{\rho_k}{\rho}(f_{,1})^2\right) \\
& + 2\sum A^2_{ml1}(f_{,1})^2 \\
\geq& (n+2)\tfrac{\rho_{,11}}{\rho}(f_{,1})^2 - C(n,\delta)\Phi(f_{,1})^2 \\
& - \delta\sum(f_{,1i})^2 + (2-\delta)\sum A^2_{ml1}(f_{,1})^2,
\end{aligned} \tag{5.9.12}$$

where δ is a positive constant as before, and $C(n,\delta)$ is a positive constant depending only on n and δ. A combination of (5.6.9) and (5.9.7) gives

$$\begin{aligned}
2\sum f_{,j}f_{,jii} \geq& (2-6\delta)\sum(f_{,1i})^2 - \tfrac{2}{\delta(n-1)}(\Delta f)^2 \\
& + (n+2)\tfrac{\rho_{,11}}{\rho}(f_{,1})^2 - C(n,\delta)\Phi(f_{,1})^2 - 2n \\
\geq& (2-6\delta)\sum(f_{,1i})^2 - \tfrac{2}{\delta(n-1)}(\Delta f)^2 - 2\delta\Psi\sum\tfrac{(\rho_{,ij})^2}{\rho^2}(f_{,1})^2 \\
& - \tfrac{(n+2)^2}{8\delta}(f_{,1})^2 - C(n,\delta)\Phi(f_{,1})^2 - 2n.
\end{aligned} \tag{5.9.13}$$

We insert (5.9.11), (5.9.13) into (5.9.8) and use (5.9.6):

$$\begin{aligned}
&\left(\tfrac{1}{2(n-1)} - 2\delta\right)\left(gf_{,1} + 2\tfrac{f_{,1}}{1+f} - \Psi_{,1}\right)^2(f_{,1})^2 - \tfrac{4}{(n-1)\delta}(\Delta f)^2 - C(n,\delta)\Phi(f_{,1})^2 \\
&\quad - \tfrac{(n+2)^2}{8\delta}(f_{,1})^2 + \Psi\sum(\Phi_{,i})^2(f_{,1})^2 - C(n,\delta,\beta)\Psi\Phi^2(f_{,1})^2 - 2n \\
&\quad + \left[-g'(f_{,1})^2 - g\Delta f + 2\tfrac{(f_{,1})^2}{(1+f)^2} - 2\tfrac{\Delta f}{1+f}\right](f_{,1})^2 \leq 0.
\end{aligned} \tag{5.9.14}$$

We choose the following values for δ and m:

$$\delta := \tfrac{1}{16(n-1)}, \qquad m := 40\,(3n + \exp\{N\})\,C,$$

where $N = \sup_{x\in\Omega}\Phi(x)$. To simplify the expression we denote

$$a_1 := \tfrac{1}{2(n-1)} - 2\delta, \quad a_2 := \tfrac{4}{(n-1)\delta},$$

$$a_3 := C(n,\delta)N + \tfrac{(n+2)^2}{8\delta} + C(n,\delta,\beta)\exp\{N\}N^2, \quad a_4 := \tfrac{a_1}{1+a_1\exp\{N\}}.$$

Recall that $\Psi_{,i} = \Psi\Phi_{,i}$. Then we have

$$a_1\left(gf_{,1} + 2\tfrac{f_{,1}}{1+f} - \Psi_{,1}\right)^2 + \Psi\sum(\Phi_{,i})^2 \geq a_4\left(gf_{,1} + 2\tfrac{f_{,1}}{1+f}\right)^2. \tag{5.9.15}$$

Inserting (5.9.15) into (5.9.14) we get

$$a_4\left(gf_{,1} + 2\tfrac{f_{,1}}{1+f}\right)^2 (f_{,1})^2 - a_2(\Delta f)^2 - a_3(f_{,1})^2$$
$$+\left(-g'(f_{,1})^2 - g\Delta f + 2\tfrac{(f_{,1})^2}{(1+f)^2} - 2\tfrac{\Delta f}{1+f}\right)(f_{,1})^2 - 2n \leq 0. \tag{5.9.16}$$

Multiply both sides of (5.9.16) by $\frac{1}{(1+f)^2}$. Then we obtain

$$a_4\left(gf_{,1} + 2\tfrac{f_{,1}}{1+f}\right)^2 \tfrac{(f_{,1})^2}{(1+f)^2} - a_2\left(\tfrac{\Delta f}{1+f}\right)^2 - a_3\tfrac{(f_{,1})^2}{(1+f)^2}$$
$$-g'\tfrac{(f_{,1})^4}{(1+f)^2} - g\Delta f\tfrac{(f_{,1})^2}{(1+f)^2} + 2\tfrac{(f_{,1})^4}{(1+f)^4} - 2\left(\tfrac{\Delta f}{1+f}\right)\tfrac{(f_{,1})^2}{(1+f)^2} - 2n \leq 0. \tag{5.9.17}$$

Using $g' \leq \frac{a_4}{20}g^2$, to further estimate (5.9.17), we have the following three inequalities:

$$g'\tfrac{(f_{,1})^4}{(1+f)^2} \leq \tfrac{a_4}{10}\left(gf_{,1} + 2\tfrac{f_{,1}}{1+f}\right)^2\tfrac{(f_{,1})^2}{(1+f)^2},$$
$$\left(\tfrac{\Delta f}{1+f}\right)^2 \leq 2n^2 + 2n^2\Phi\tfrac{(f_{,1})^2}{(1+f)^2},$$
$$2\left(\tfrac{\Delta f}{1+f}\right)\tfrac{(f_{,1})^2}{(1+f)^2} \leq 2n\tfrac{(f_{,1})^2}{(1+f)^2} + \tfrac{(f_{,1})^4}{(1+f)^4} + n^2\Phi\tfrac{(f_{,1})^2}{(1+f)^2}.$$

We use the inequality of Schwarz and obtain

$$g\Delta f\tfrac{(f_{,1})^2}{(1+f)^2} = ng\tfrac{(f_{,1})^2}{(1+f)^2} + \tfrac{n+2}{2}g\sum\tfrac{\rho_{,i}}{\rho}f_{,i}\tfrac{(f_{,1})^2}{(1+f)^2}$$
$$\leq \tfrac{5n^2}{2a_4} + \tfrac{a_4}{10}g^2\tfrac{(f_{,1})^4}{(1+f)^4} + \tfrac{10(n+2)^2}{16a_4}\Phi\tfrac{(f_{,1})^2}{(1+f)^2} + \tfrac{a_4}{10}g^2\tfrac{(f_{,1})^4}{(1+f)^2}$$
$$\leq \tfrac{a_4}{5}\left(gf_{,1} + 2\tfrac{f_{,1}}{1+f}\right)^2\tfrac{(f_{,1})^2}{(1+f)^2} + \tfrac{5n^2}{2a_4} + \tfrac{5n^2}{2a_4}\Phi\tfrac{(f_{,1})^2}{(1+f)^2}.$$

We insert these inequalities into (5.9.17) and get

$$\tfrac{(f_{,1})^4}{(1+f)^4} - a\tfrac{(f_{,1})^2}{(1+f)^2} - b \leq 0, \tag{5.9.18}$$

where we use the abbreviations:

$$a := \left(2n^2a_2 + \tfrac{5n^2}{2a_4} + n^2\right)N + 2n + a_3,$$
$$b := 2n^2a_2 + \tfrac{5n^2}{2a_4} + 2n.$$

The left hand term in (5.9.18) is a quadratic expression in $\frac{(f_{,1})^2}{(1+f)^2}$. If one considers its zeroes it follows that

$$\tfrac{(f_{,1})^2}{(1+f)^2} \leq a + \sqrt{b}.$$

Thus from (5.9.1) we get, with our special choice of δ and m:

$$F \leq \exp\{\exp\{N\}\}(a + \sqrt{b}),$$

which holds at p^*, where F attains its supremum. Hence, at any interior point of $S_f(0, C)$, we have

$$\frac{\|\text{grad } f\|^2}{(1+f)^2} \leq \exp\{\exp\{N\}\}(a+\sqrt{b})\exp\left\{\frac{40(3n+\exp\{N\})C}{C-f}\right\}. \tag{5.9.19}$$

Let $C \to \infty$ then

$$\frac{\|\text{grad } f\|^2}{(1+f)^2} \leq \exp\{\exp\{N\} + 40(3n + \exp\{N\})\}(a+\sqrt{b}) := Q, \tag{5.9.20}$$

where Q is a constant.
Using the gradient estimate (5.9.20) we can prove that M is complete with respect to the Calabi metric, namely: for any unit speed geodesic, starting from p,

$$\sigma : [0, S] \to M$$

we have

$$\frac{df}{ds} \leq \|\text{grad } f\| \leq \sqrt{Q}(1+f).$$

It follows that

$$s \geq \frac{1}{\sqrt{Q}} \int_0^{x_{n+1}(\sigma(S))} \frac{df}{1+f}. \tag{5.9.21}$$

Since

$$\int_0^{\infty} \frac{df}{1+f} = \infty$$

and $f : \Omega \to \mathbb{R}$ is proper (i.e., the inverse image of any compact set is compact), (5.9.21) implies that M is complete with respect to the Calabi metric. This completes the proof of Theorem 5.9.1. ■

Chapter 6

Hypersurfaces with Constant Affine Mean Curvature

6.1 Classification

The classification of locally strongly convex, affine-complete affine hyperspheres had attracted many geometers since about 1920. For the history of this problem and the contributions of different authors we refer to the monograph [58], pp. 84-85. Obviously, every affine hypersphere has constant affine mean curvature L_1. Thus the next interesting and important problem is the classification of locally strongly convex, complete affine hypersurfaces with constant affine mean curvature L_1. We shall show that the study of locally strongly convex, Euclidean complete hypersurfaces with constant affine mean curvature $L := L_1$ is equivalent to the study of the convex solutions of the fourth order PDE

$$\Delta \left[\det\left(\frac{\partial^2 f}{\partial x^i \partial x^j}\right)\right]^{\frac{-1}{n+2}} = -nL \left[\det\left(\frac{\partial^2 f}{\partial x^i \partial x^j}\right)\right]^{\frac{-1}{n+2}}, \tag{6.1.1}$$

where Δ denotes the Laplacian with respect to the Blaschke metric. Here, we consider the convex solutions of the equation (6.1.1) for $f : \mathbb{R}^2 \to \mathbb{R}$. We shall prove that

a) if $L > 0$, then there is no convex solution of (6.1.1) which is defined for all $(x^1, x^2) \in \mathbb{R}^2$;

b) if $L = 0$ and $f(x^1, x^2)$ is a convex solution of (6.1.1), which is defined for all $(x^1, x^2) \in \mathbb{R}^2$, then $f(x^1, x^2)$ must be a quadratic polynomial.

In the language of affine differential geometry, we are going to prove the following theorems:

Theorem 6.1.1. [53]. *Every locally strongly convex, Euclidean complete surface with constant affine mean curvature is affine complete.*

To state our Theorem 6.1.2 we introduce the following terminology. A locally strongly convex hypersurface is called to have *finite geometry*, if $\|B\|_k$ and $\|A\|_k$

are bounded, where

$$\|B\|_k := \|B\| + \|\nabla B\| + \|\nabla^2 B\| + \cdots + \|\nabla^k B\|,$$
$$\|A\|_k := \|A\| + \|\nabla A\| + \|\nabla^2 A\| + \cdots + \|\nabla^k A\|,$$

and where, according to chapter 2, B denotes the Weingarten form and A the cubic form. The norm $\|\cdot\|$ and the covariant derivative ∇ are defined with respect to the Blaschke metric, and $\nabla^k := \nabla(\nabla^{k-1})$.

Using Theorem 6.1.1 we immediately get the following result (see [53]):

Theorem 6.1.2. *Let M be a locally strongly convex, Euclidean complete surface in $\mathbb{R}^3$ with constant affine mean curvature L.*

(a) *If $L > 0$ then M is an ellipsoid.*

(b) *If $L = 0$ then M is an elliptic paraboloid.*

(c) *If $L < 0$ then M has "finite geometry".*

As a corollary of Theorem 6.1.2, we present a new proof of Chern's conjecture about affine maximal surfaces. We state the following conjecture for higher dimension:

Conjecture 6.1.3. *Let M be a locally strongly convex, Euclidean complete hypersurface in $\mathbb{R}^{n+1}$ with constant affine mean curvature L.*

(a) *If $L > 0$ then M is an ellipsoid.*

(b) *If $L = 0$ then M is an elliptic paraboloid.*

(c) *If $L < 0$ then M has "finite geometry".*

6.1.1 *Estimates for the determinant of the Hessian*

As in section 5.6.2 we prove two lemmas in order to estimate the determinant of the Hessian of certain functions from above. We use the definition of ρ in sections 1.4 and 3.3.4, and formula (3.3.9). In terms of the Calabi metric the PDE (6.1.1) can be rewritten as

$$\Delta\rho = \tfrac{2-n}{2}\tfrac{\|\text{grad}\,\rho\|^2}{\rho} - nL\rho^2. \tag{6.1.2}$$

In particular, when $n = 2$, the PDE (6.1.2) reduces to

$$\Delta\rho = -2L\rho^2.$$

Note that, when $a = -\frac{3}{4}$, the PDE (5.6.3) reduces to

$$\Delta\rho = -\tfrac{1}{3}L^\sharp\rho^2.$$

Both equations are of the same type of PDEs. Thus, for a constant affine mean curvature surface, we can use Lemmas 5.6.6 and 5.6.7 to obtain the estimates for the determinant:

Lemma 6.1.4. *Let Ω be a bounded convex domain with $(0,0) \in \Omega$, and f be a strictly convex function defined on Ω satisfying the PDE (6.1.1). Let u be the Legendre transform function of f and Ω'^* be an arbitrary subdomain of the Legendre transform domain Ω^* such that $dist(\Omega'^*, \partial\Omega^*) > 0$. Then the following estimate holds:*

$$\det(u_{ij}) \leq C_1, \qquad for\ \xi \in \Omega'^*,$$

where C_1 is a constant depending only on $dist(\Omega'^*, \partial\Omega^*)$, $diam(\Omega)$, $diam(\Omega^*)$ and $|L|$.

Lemma 6.1.5. *Let Ω be a bounded convex domain, and f be a strictly convex function defined on Ω satisfying the PDE (6.1.1). Let Ω' be an arbitrary subdomain of Ω with $dist(\Omega', \partial\Omega) > 0$. Then the following estimate holds:*

$$\det(f_{ij}) \leq C_2, \qquad for\ x \in \Omega',$$

where C_2 is a constant depending only on $dist(\Omega', \partial\Omega)$, $diam(\Omega)$, $diam(\Omega^*)$ and $|L|$.

6.1.2 *Proof of Theorem 6.1.1*

Suppose that M is a locally strongly convex, Euclidean complete affine surface. Obviously, M is affine complete if M is compact. Therefore, it is enough to consider the case when M is a non-compact, Euclidean complete, locally strongly convex surface with constant affine mean curvature L. From Hadamard's Theorem (see section 1.2.1) M is the graph of a strictly convex function $x^3 = f(x^1, x^2)$ defined in a convex domain $V \subset \mathbb{R}^2$. To prove Theorem 6.1.1, we need the completeness criterion, see section 4.2.2.

Now we use arguments from *blow-up analysis* to show that there is a constant $N > 0$ such that the Weingarten form satisfies $\|B\|_G^2 \leq N$ everywhere; the proof of this upper bound will take all of subsection 6.1.2. Then Theorem 6.1.1 follows from the completeness criterion.
To this end, assume that $\|B\|_G^2$ is unbounded. Then there exists a sequence of points $\{p_k\} \subset M$ such that

$$\|B\|_G^2(p_k) \to \infty$$

as $k \to \infty$. For each $p_k \in M$ we may assume that the plane $x^3 = 0$ is the tangent plane of M at p_k and p_k has the coordinates $(0,0)$. With respect to this coordinate system, we have $f \geq 0$, and for any real number $C > 0$ the section

$$S_f(0, C) = \{(x^1, x^2) \in V \mid x^3 = f(x^1, x^2) < C\}$$

is a bounded convex domain in $\mathbb{R}^2$. As already stated several times (see [37], p.27), there exists a unique ellipsoid E, which attains the minimum volume among all ellipsoids that contain $S_f(0, C)$ and that are centered at the center of mass of $S_f(0, C)$ such that

$$\tfrac{1}{2\sqrt{2}} E \subset S_f(0, C) \subset E, \tag{6.1.3}$$

where $\frac{1}{2\sqrt{2}}E$ means the $\frac{1}{2\sqrt{2}}$-dilation of E with respect to its center. By an orthogonal linear transformation, we may assume that the equation of the minimum ellipsoid E is

$$\frac{(x^1-\dot{x}^1)^2}{a_1^2}+\frac{(x^2-\dot{x}^2)^2}{a_2^2}=1,$$

where $\dot{x}$ is the center of mass of $S_f(0,C)$. By the following unimodular affine transformation

$$\widehat{x}^1=\sqrt{C\cdot\tfrac{a_2}{a_1}}\cdot x^1,\qquad \widehat{x}^2=\sqrt{C\cdot\tfrac{a_1}{a_2}}\cdot x^2,\qquad \widehat{x}^3=\tfrac{1}{C}\cdot x^3, \tag{6.1.4}$$

M is given as a graph of a strictly convex function $\widehat{f}(\widehat{x}^1,\widehat{x}^2)$ defined on a convex domain $\widehat{\Omega}\subset\mathbb{R}^2$. Denote by L_C the linear transformation

$$\widehat{x}^1=\sqrt{C\cdot\tfrac{a_2}{a_1}}\cdot x^1,\qquad \widehat{x}^2=\sqrt{C\cdot\tfrac{a_1}{a_2}}\cdot x^2.$$

Then, $L_C(E)$ is the ball B_δ with center $\left(\sqrt{C\cdot\frac{a_2}{a_1}}\cdot\dot{x}^1,\sqrt{C\cdot\frac{a_1}{a_2}}\cdot\dot{x}^2\right)$ and radius $\delta=\sqrt{Ca_1a_2}$. Setting $\Omega_C=L_C(S_f(0,C))$, (6.1.3) becomes

$$\frac{1}{2\sqrt{2}}B_\delta\subset\Omega_C\subset B_\delta. \tag{6.1.5}$$

Obviously, we have

$$\Omega_C=\left\{(\widehat{x}^1,\widehat{x}^2)\in\widehat{\Omega}\mid \widehat{f}(\widehat{x}^1,\widehat{x}^2)<1\right\}.$$

It is easy to see that the function

$$\mathbb{V}:(0,\infty)\to\mathbb{R},\quad \mathbb{V}(C)=\pi\cdot C\cdot a_1(C)a_2(C)$$

is continuous. Note that $\mathbb{V}((0,\infty))=(0,\infty)$. It follows that there exists a number $C^{(k)}>0$, such that $C^{(k)}a_1(C^{(k)})a_2(C^{(k)})=1$. This implies that, by a unimodular affine transformation (6.1.4) with $C=C^{(k)}$, M is given by a strictly convex function $f^{(k)}$ defined in a convex domain in $\mathbb{R}^2$ such that

$$B_{\frac{1}{2\sqrt{2}}}(x^{(k)})\subset\Omega_k\subset B_1(x^{(k)}),$$

where

$$\Omega_k:=\left\{(x^1,x^2)\mid f^{(k)}<1\right\}.$$

Thus, we would obtain a sequence of convex functions $\{f^{(k)}\}$ and a sequence of points $\{x^{(k)}\}$ such that $f^{(k)}\geq 0$, and such that $B_{\frac{1}{2\sqrt{2}}}(x^{(k)})\subset\Omega_k\subset B_1(x^{(k)})$. Therefore, we may assume, by taking subsequences, that $\{\Omega_k\}$ converges to a convex domain Ω and $\{f^{(k)}\}$ converges to a convex function f_∞, locally uniformly in Ω. For $x\in\partial\Omega$, we define $f_\infty(x)=\underline{\lim}_{y\to x,y\in\Omega}f_\infty(y)$.

In the following we shall give uniform estimates of $\det\left(\frac{\partial^2u^{(k)}}{\partial\xi_i\partial\xi_j}\right)$ from below and above (where $u^{(k)}$ denotes the Legendre transformation relative to $f^{(k)}$), and use the Caffarelli-Gutierrez theory to obtain a Hölder estimate for $\det\left(\frac{\partial^2u^{(k)}}{\partial\xi_i\partial\xi_j}\right)$ (see [15] or [91], Theorem 4.1). Then we use the Caffarelli-Schauder estimate for the Monge-Ampère equation [14] to get a $C^{2,\alpha}$ estimate, to show that the limit surface is a smooth surface.

I. Estimate for $\det\left(\frac{\partial^2 u^{(k)}}{\partial \xi_i \partial \xi_j}\right)$ **from below.**

Let us denote

$$D := \{x \in \overline{\Omega} \mid f_\infty(x) = 0\},$$

where $\overline{\Omega}$ denotes the closure of Ω. It is easy to see that D is a closed subset of $\overline{\Omega}$ and $(0,0) \in D$. To estimate $\det\left(\frac{\partial^2 u^{(k)}}{\partial \xi_i \partial \xi_j}\right)$, we shall consider different cases according to the location of D:

Case 1: $D \subset \Omega$.

Case 2: $D \cap \partial\Omega \neq \emptyset$.

Our aim is to show that the Case 2 cannot take place.
We consider Case 1. In this case, there exists a number b, $0 < b < 1$, such that the set

$$\overline{\Omega_{2b}} \subset \Omega,$$

where $\Omega_{2b} := \left\{(x^1, x^2) \in \Omega \mid f_\infty(x^1, x^2) < 2b\right\}$. Put

$$\Omega_{k,b} := \left\{(x^1, x^2) \in \Omega_k \mid f^{(k)}(x^1, x^2) < b\right\}.$$

Since $\{f^{(k)}\}$ converges to f_∞ locally uniformly, we have

$$\overline{\Omega_{k,b}} \subset \overline{\Omega_{2b}} \subset \Omega$$

for k large enough. It follows that

$$\operatorname{dist}\left(\Omega_{k,b}, \partial\Omega_k\right) > d$$

for k large enough, where $d > 0$ is a constant independent of k. Now we use Lemma 6.1.5 to conclude that

$$\det\left(\frac{\partial^2 f^{(k)}}{\partial x^i \partial x^j}\right) \leq d_1, \quad \text{for } x \in \Omega_{k,b}, \tag{6.1.6}$$

where $d_1 > 0$ is a constant depending only on $|L|$ and d. Consider the Legendre transformation relative to $f^{(k)}$:

$$\xi_i^{(k)} = \frac{\partial f^{(k)}}{\partial x^i}, \quad u^{(k)} = \sum x^i \frac{\partial f^{(k)}}{\partial x^i} - f^{(k)}.$$

Set

$$\Omega_k^* = \left\{(\xi_1^{(k)}(x), \xi_2^{(k)}(x)) \mid (x^1, x^2) \in \Omega_k\right\},$$

$$\Omega_{k,b}^* = \left\{(\xi_1^{(k)}(x), \xi_2^{(k)}(x)) \mid (x^1, x^2) \in \Omega_{k,b}\right\}.$$

Then, by (6.1.6), we have

$$\det\left(\frac{\partial^2 u^{(k)}}{\partial \xi_i \partial \xi_j}\right) \geq \frac{1}{d_1}, \quad \text{for } \xi \in \Omega_{k,b}^*. \tag{6.1.7}$$

II. Estimate for $\det\left(\frac{\partial^2 u^{(k)}}{\partial\xi_i\partial\xi_j}\right)$ **from above.**

To get the estimate, we need some important results from the classical theory of convex bodies in Euclidean space (see [5], [13]). Let F be a convex hypersurface in $\mathbb{R}^{n+1}$ and e be a subset of F. We denote by $\psi_F(e)$ the Euclidean *spherical image* of e. If the set e is a Borel set, the spherical image of the set e is also a Borel set and therefore it is measurable. Denote by $\sigma_F(e)$ the *area (measure)* of the spherical image $\psi_F(e)$ of the Borel set e of F and call it the *integral Gaußian curvature* of e. Denote by $A(e)$ the measure (or area) of the Borel set e on F. The ratio $\frac{\sigma_F(e)}{A(e)}$ is called the *specific curvature* of e. The following theorems hold (see [5], or [13], p. 35):

Theorem 6.1.6. (A.D. Aleksandrov). *A convex surface whose specific curvature is bounded away from zero is strictly convex.*

Theorem 6.1.7. *Let a sequence of closed convex hypersurfaces F_k converge to a closed convex hypersurface F and a sequence of closed subset M_k of F_k converge to a closed subset M of F; then*

$$\sigma_F(M) \geq \overline{\lim}_{k\to\infty}\sigma_{F_k}(M_k).$$

Claim: First of all, we claim that there exists a ball $B^*_\delta(0)$ such that

$$B^*_\delta(0) \subset \Omega^*_{k,b}, \qquad \text{for} \quad k = 1, 2, \cdots$$

In fact, since Ω is bounded, there is a ball $B_{r_1}(0)$ with the center $(0,0)$ and the radius r_1 such that

$$\overline{\Omega_b} := \{(x^1,x^2) \in \Omega \mid f_\infty(x^1,x^2) \leq b\} \subset B_{r_1}(0).$$

Since $\{f^{(k)}\}$ converges to f_∞ locally uniformly in Ω, we see that

$$\overline{\Omega_{k,\frac{b}{2}}} := \left\{(x^1,x^2) \in \Omega_k \mid f^{(k)}(x^1,x^2) \leq \tfrac{b}{2}\right\} \subset B_{r_1}(0)$$

for k large enough. Consider the convex cone K with vertex $(0,0)$ and the base

$$\left\{(x^1,x^2,\tfrac{b}{2}) \mid (x^1,x^2) \in \partial\Omega_{k,\frac{b}{2}}\right\}.$$

Then we have the normal mapping relation (see section 1.3)

$$\partial f^{(k)}\left(\Omega_{k,\frac{b}{2}}\right) \supset \partial K\left(\Omega_{k,\frac{b}{2}}\right).$$

On the other hand, since

$$\Omega_{k,\frac{b}{2}} \subset B_{r_1}(0),$$

we see that (see [5], p.126, or [37], p.1)

$$\partial K\left(\Omega_{k,\frac{b}{2}}\right) \supset B^*_{\frac{b}{2r_1}}(0)$$

and the claim follows.

Next, we want to prove that u_∞ is strictly convex at (0, 0).
In fact, since the functions $u^{(k)}$ are convex and bounded, we may assume, by taking subsequences, that $\{u^{(k)}\}$ converges to a convex function u_∞, locally uniformly in $\Omega^*_{\frac{b}{2r_1}} = \left\{(\xi_1,\xi_2) \mid \sum(\xi_i)^2 < \frac{b^2}{4r_1^2}\right\}$. Let e be a closed subset of $\Omega^*_{\frac{b}{2r_1}}$ with $e^o \neq \emptyset$, where e^o denotes the interior of e. By F and $F^{(k)}$ we denote the graphs of the functions $u_\infty : \Omega^*_{\frac{b}{2r_1}} \to \mathbb{R}$ and $u^{(k)} : \Omega^*_{\frac{b}{2r_1}} \to \mathbb{R}$, respectively.

Set

$$F_e = \{(\xi_1,\xi_2,u_\infty(\xi_1,\xi_2)) \mid (\xi_1,\xi_2) \in e\},$$

$$F_e^{(k)} = \left\{\left(\xi_1,\xi_2,u^{(k)}(\xi_1,\xi_2)\right) \mid (\xi_1,\xi_2) \in e\right\}.$$

Then, by Theorem 6.1.7 and (6.1.7), we get

$$\begin{aligned}\sigma_F(F_e) &\geq \overline{\lim}_{k\to\infty}\sigma_{F^{(k)}}(F_e^{(k)})\\ &= \overline{\lim}_{k\to\infty}\int_{F_e^{(k)}} \frac{\det\left(\frac{\partial^2 u^{(k)}}{\partial\xi_i\partial\xi_j}\right)}{\left(1+\sum\left(\frac{\partial u^{(k)}}{\partial\xi_i}\right)^2\right)^2} dp\\ &\geq b_2\overline{\lim}_{k\to\infty}A(F_e^{(k)})\\ &= b_2 A(F_e),\end{aligned}$$

where b_2 is a constant depending only d_1 and diam(Ω), i.e.,

$$\frac{\sigma_F(F_e)}{A(F_e)} \geq b_2 > 0. \tag{6.1.8}$$

We apply Theorem 6.1.6 and conclude that u_∞ is strictly convex at (0,0).

Now we are ready to estimate $\det\left(\frac{\partial^2 u^{(k)}}{\partial\xi_i\partial\xi_j}\right)$ from above. Since u_∞ is strictly convex at $(0,0)$, there exists a positive constant $0 < h_1 < 1$, such that

$$\Omega^*_{h_1} = \left\{(\xi_1,\xi_2) \in \Omega^*_{\frac{b}{2r_1}} \mid u_\infty(\xi_1,\xi_2) < h_1\right\}$$

is a bounded convex domain. Then we choose $0 < h_2 < h_1$ such that $\overline{\Omega^*_{2h_2}} \subset \Omega^*_{h_1}$, where

$$\Omega^*_{2h_2} = \{(\xi_1,\xi_2) \in \Omega^*_{h_1} \mid u_\infty(\xi_1,\xi_2) < 2h_2\}.$$

Put

$$\Omega^*_{k,h_2} := \left\{(\xi_1,\xi_2) \in \Omega^*_k \mid u^{(k)}(\xi_1,\xi_2) < h_2\right\}.$$

Since $u^{(k)}$ converges to u_∞ locally uniformly, we have $\overline{\Omega^*_{k,h_2}} \subset \Omega^*_{h_1}$, and there exists a constant $d_2 > 0$ such that

$$\operatorname{dist}\left(\Omega^*_{k,h_2}, \partial\Omega^*_{h_1}\right) > d_2$$

for k large enough. Clearly, there is a uniform estimate

$$\sum \left(\frac{\partial u^{(k)}}{\partial \xi_i} \right)^2 \leq \text{diam}(\Omega), \qquad \text{for } \xi \in \Omega^*_{\frac{b}{2r_1}}.$$

Now we use Lemma 6.1.4 to conclude that

$$\det \left(\frac{\partial^2 u^{(k)}}{\partial \xi_i \partial \xi_j} \right) < d_3, \qquad \text{for } \xi \in \Omega^*_{k,h_2}, \tag{6.1.9}$$

for k large enough, where $d_3 > 0$ is a constant depending only on diam(Ω), d_2 and $|L|$. (6.1.9) gives the upper bound. ■

Now we are ready to prove the following lemma:

Lemma 6.1.8. *There exists a neighborhood U of (0,0) such that*

$$D^2 u^{(k)} \geq C_3 I, \quad |D^l u^{(k)}| \leq C_4, \qquad \textit{for } k, l = 1, 2, \cdots, \tag{6.1.10}$$

where C_3 and C_4 are constants. C_3 depends only on d_2, diam(Ω), $|L|$ and d, while C_4 depends additionally on l.

Proof. We set

$$v = \frac{1}{\rho}, \quad \rho_i = \frac{\partial \rho}{\partial \xi_i}, \quad v_i = \frac{\partial v}{\partial \xi_i} \quad v_{ij} = \frac{\partial^2 v}{\partial \xi_i \partial \xi_j}.$$

Then, by (2.7.8), in terms of the Blaschke metric, we get:

$$\Delta \left(\tfrac{1}{v} \right) = \tfrac{1}{\rho} \sum u^{ij} \left(-\tfrac{v_{ij}}{v^2} + 2 \tfrac{v_i v_j}{v^3} \right) - \tfrac{2}{\rho^2} \sum u^{ij} \tfrac{v_i v_j}{v^4} = -\rho \sum u^{ij} v_{ij}.$$

On the other hand, by (2.7.2), we have

$$\Delta \left(\tfrac{1}{v} \right) = \Delta \rho = -nL\rho$$

and it follows that

$$\sum u^{ij} v_{ij} - nL = 0.$$

Therefore, setting

$$\Psi := v - Lu + 2|L|,$$

we obtain

$$\sum u^{ij} \Psi_{ij} = 0, \tag{6.1.11}$$

where $\left(u^{ij}\right)$ is the inverse matrix of (u_{ij}). By (6.1.7), (6.1.9) and (6.1.11), we may use the Caffarelli-Gutierrez theory to obtain a Hölder estimate for $\det \left(\frac{\partial^2 u^{(k)}}{\partial \xi_i \partial \xi_j} \right)$ (see [15] or [91], Theorem 4.1). Then, to get a $C^{2,\alpha}$ estimate, we use the Caffarelli-Schauder estimate for the Monge-Ampère equation [14]. Finally, by bootstrapping, Lemma 6.1.8 follows. ■

Consequently, from Lemma 6.1.8 it follows that u_∞ is a smooth strictly convex function in a neighborhood of (0,0), and hence f_∞ is a smooth strictly convex function in a neighborhood of (0,0).

Now it is our purpose to show that Case 2 in subsection (**I**) cannot take place. Let

$$V_k := \left\{(x^1,x^2,x^3)\in\mathbb{R}^3 \mid f^{(k)}(x^1,x^2)\le x^3\le 1,\ (x^1,x^2)\in\overline{\Omega_k}\right\},$$

$$V_\infty := \left\{(x^1,x^2,x^3)\in\mathbb{R}^3 \mid f_\infty(x^1,x^2)\le x^3\le 1,\ (x^1,x^2)\in\overline{\Omega}\right\}.$$

Then the sequence of convex bodies $\{V_k\}$ converges to the convex body V_∞.

Claim: The set D, *defined in (*__I__*) above, is a line segment or a single point.*

To prove this claim, first we show that there exists a ball $B^*_{r_0}(0)$ with center (0,0) and the radius r_0 such that

$$B^*_{r_0}(0)\subset\Omega^*_k \qquad \text{for } k=1,2,\cdots$$

As before, we choose a ball $B_l(0)$ with center (0,0) and radius l such that $\Omega\subset B_l(0)$. Since Ω_k converges to Ω, we see that

$$\Omega_k\subset B_l(0)$$

for k large enough. Then it is easy to see that $\Omega^*_k\supset B^*_{\frac{1}{l}}(0)$ for $k=1,2,\cdots$

Now we prove our claim. By contradiction let us assume that there exists a ball

$$B_\epsilon(x_0)=\left\{(x^1,x^2)\mid \sum(x^i-(x_0)^i)^2<\epsilon^2\right\},$$

such that $B_\epsilon(x_0)\subset D$. Since $\{f^{(k)}\}$ converges to f_∞ locally uniformly, there is a positive number k_0, such that

$$0\le f^{(k)}(x)<\tfrac{r_0\epsilon}{8},\qquad \text{for}\ \ x\in B_{\frac{\epsilon}{2}}(x_0),\quad (k>k_0).$$

Clearly, there exists a uniform estimate

$$\sum\left(\tfrac{\partial f^{(k)}}{\partial x^i}\right)^2<\tfrac{r_0^2}{4},\qquad \text{for}\ \ x\in B_{\frac{\epsilon}{4}}(x_0),\quad (k>k_0).$$

Put

$$\Omega^*_{k,\frac{\epsilon}{4}}=\left\{(\xi_1^{(k)}(x),\xi_2^{(k)}(x))\mid x\in B_{\frac{\epsilon}{4}}(x_0)\right\}.$$

Then we have

$$\Omega^*_{k,\frac{\epsilon}{4}}\subset B^*_{\frac{r_0}{2}}(0)\subset\Omega^*_k,\qquad \text{for}\quad k>k_0.$$

Note that $B_{r_0}(0)^*\subset\Omega^*_k$ for $k=1,2,\cdots$. Hence we use Lemma 6.1.4 to conclude that there exists a constant $d_4>0$, depending only on $r_0,\operatorname{diam}(\Omega)$ and $|L|$, such that

$$\det\left(\tfrac{\partial^2u^{(k)}}{\partial\xi_i\partial\xi_j}\right)<d_4\qquad \text{for } \xi\in B^*_{\frac{r_0}{2}}(0).$$

This implies that

$$\det\left(\tfrac{\partial^2 f^{(k)}}{\partial x^i\partial x^j}\right)>\tfrac{1}{d_4},\qquad \text{for}\ \ x\in B_{\frac{\epsilon}{4}}(x_0),\ \ (k>k_0). \tag{6.1.12}$$

Therefore we can apply the argument of Case 1 to $\{f^{(k)}\}$ and conclude that the function f_∞ is strictly convex at x_0. This contradiction shows that D must be a line segment or a single point.

Now we are ready to prove that Case 2 cannot take place. We shall consider the following two cases:

Case 2.1. $D \cap \partial\Omega$ contains at most two points.

Case 2.2. D is a line segment with $D \subset \partial\Omega$.

Case 2.1. Let $p \in D \cap \partial\Omega$ and let l be a supporting line of Ω at p. The line l and the unit normal ν of the (x^1, x^2)-plane determine a plane. We denote this plane by P. Then the plane P and the (x^1, x^2)-plane divide the space $\mathbb{R}^3$ into four closed subspaces such that V_∞ lies completely in one of them. Let α be a supporting plane of V_∞ containing the line l such that it intersects P and forms an angle $\angle(\alpha, P) = \theta$ with P, where $\theta > 0$ is sufficiently small (see Figure 1). Since $p \in \partial V_\infty$ and α

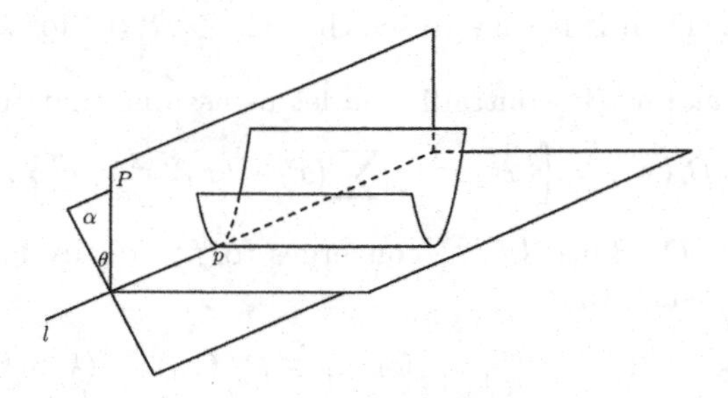

Figure 1.

is a supporting plane of V_∞, there is a neighborhood $U \subset \partial V_\infty$, which projects orthogonally and one-to-one onto a convex domain $\Omega^{(1)} \subset \alpha$. This implies that, near to the point p, ∂V_∞ can be represented as the graph of a convex function g defined in $\Omega^{(1)}$. Obviously, g is strictly convex at p but it is not necessarily smooth at p. We choose a number b, $0 < b < 1$, such that

$$\Omega^{(2)} = \left\{(y_1, y_2) \in \Omega^{(1)} \mid g(y_1, y_2) < b\right\}$$

is a bounded convex domain in $\mathbb{R}^2$. Then we choose a new coordinate system $\{y_1, y_2, y_3\}$ such that

1) p has coordinates $(0, 0, 1)$.

2) The equation of α is $y_3 = 1$.

Since the sequence of convex bodies $\{V_k\}$ converges to the convex body V_∞, we see that the boundary ∂V_k of V_k can also be represented as the graph of a convex function $g^{(k)}$ for sufficiently large k. Obviously, $g^{(k)} \to g + 1$ in a bounded convex domain $\Omega^{(3)}$. Note that the graph of $g^{(k)}$ is a locally strongly convex surface with constant affine mean curvature L. We can therefore apply the argument of Case 1 to $\{g^{(k)}\}$ and conclude that the function g is a smooth function near the point p. The contradiction shows that Case 2.1 cannot take place.

Case 2.2. In this case, we have $p = (0,0) \in \partial\Omega$. Let l be the line containing D. We choose a new coordinate system $\{y_1, y_2, y_3\}$ as in Case 2.1 (see Figure 2). Then, the

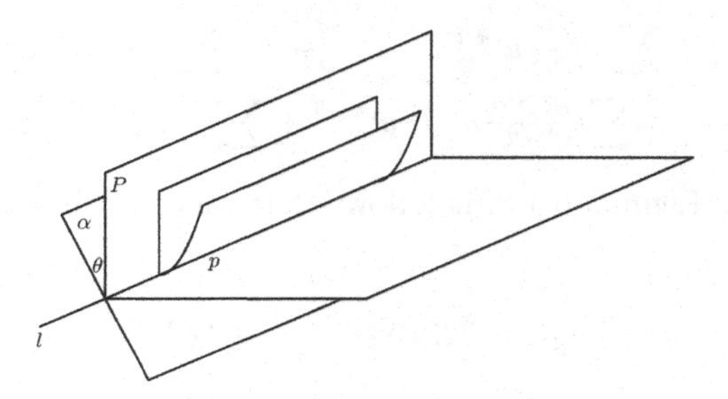

Figure 2.

boundary ∂V_∞ of V_∞ can be represented as the graph of a convex function g defined on a convex domain $\Omega^{(3)}$. With respect to this coordinate system we have $g \geq 1$. The boundary ∂V_k of V_k can also be represented as a graph of a convex function $g^{(k)}$ for sufficiently large k. Obviously, $g^{(k)} \to g$ in $\Omega^{(3)}$. Note that the graph of $g^{(k)}$ is a locally strongly convex surface with constant affine mean curvature L. Again, we shall consider different cases according to the location of

$$D^\sharp := \left\{ (y_1, y_2) \in \overline{\Omega}^{(3)} \mid g(y_1, y_2) = 1 \right\}.$$

Case 2.2.1 $D^\sharp \subset \Omega^{(3)}$.

Case 2.2.2 $D^\sharp \cap \partial\Omega^{(3)} \neq \emptyset$.

Case 2.2.1 Note that $D^\sharp$ is a line segment. Since $D^\sharp \subset \Omega^{(3)}$, we can apply the argument of Case 1 to Case 2.2.1 and conclude that Case 2.2.1 cannot take place.

Case 2.2.2 In this case, $D^\sharp \cap \partial\Omega^{(3)}$ contains at most two points. Therefore we can apply the argument of Case 2.1 to Case 2.2.2 and conclude that Case 2.2.2 cannot take place.

Now we are in a position to prove that $\|B\|_G^2$ is bounded. By Lemma 6.1.8, we have

$$D^2 u^{(k)} \geq C_3 I, \qquad |D^l u^{(k)}| \leq C_4, \quad l = 0, 1, 2, \cdots$$

in a neighborhood U of (0,0), where C_3 and C_4 are constants. C_3 depends only on $d, d_2, \mathrm{diam}(\Omega)$ and $|L|$, and C_4 additionally depends on l. Note that $\|B\|_G^2$ is equiaffinely invariant. By (2.7.6), we have

$$\|B\|_G^2(p_k) = \sum G^{is}_{(k)} G^{jt}_{(k)} B^{(k)}_{ij} B^{(k)}_{st} \, |_{(0,0)} \,,$$

where

$$B_{ij}^{(k)} = -\frac{1}{\rho^{(k)}}\frac{\partial^2\rho^{(k)}}{\partial x^i \partial x^j} + \frac{2}{(\rho^{(k)})^2}\frac{\partial \rho^{(k)}}{\partial x^i}\frac{\partial \rho^{(k)}}{\partial x^j} + \sum \frac{f^{(k)st}}{\rho^{(k)}}\frac{\partial \rho^{(k)}}{\partial x^s}\frac{\partial f_{ij}^{(k)}}{\partial x^t},$$

$$\rho^{(k)} = \left[\det\left(\frac{\partial^2 f^{(k)}}{\partial x^i \partial x^j}\right)\right]^{-\frac{1}{4}}, \qquad G_{ij}^{(k)} = \rho^{(k)}\frac{\partial^2 f^{(k)}}{\partial x^i \partial x^j}.$$

Note that

$$\frac{\partial \rho^{(k)}}{\partial x^i} = \sum \frac{\partial \rho^{(k)}}{\partial \xi_l} u^{(k)li}, \qquad \frac{\partial f_{ij}^{(k)}}{\partial x^l} = \sum \frac{\partial u^{(k)ij}}{\partial \xi_s} u^{(k)sl},$$

$$\frac{\partial^2 \rho^{(k)}}{\partial x^i \partial x^j} = \sum \frac{\partial^2 \rho^{(k)}}{\partial \xi_l \partial \xi_s} u^{(k)li} u^{(k)sj} + \sum \frac{\partial \rho^{(k)}}{\partial \xi_l}\frac{\partial u^{(k)li}}{\partial \xi_s} u^{(k)sj}.$$

Consequently, from Lemma 6.1.8, it follows that there exists a number $N > 0$ such that

$$\|B\|_G^2(p_k) \leq N, \qquad k = 1, 2, \cdots$$

On the other hand, we have

$$\|B\|_G^2(p_k) \to \infty, \quad \text{as} \quad k \to \infty.$$

The contradiction shows that there must exist a number $N > 0$, such that $\|B\|_G^2 \leq N$ on M. Then, by the completeness criterion, Theorem 6.1.1 follows. ■

6.1.3 *Proof of Theorem 6.1.2*

Recall that there is no locally strongly convex, compact hypersurface without boundary and with non-positive affine mean curvature (see [58], p.121). This implies that M is non-compact if M is a complete, locally strongly convex surface with constant affine mean curvature $L \leq 0$; this concerns the cases **(b)** and **(c)** below.

(a) Denote by R the scalar curvature, we have $R = 2(J + L) \geq 2L > 0$ (see (2.5.8)). Moreover, if M is Euclidean complete, by Theorem 6.1.1, M is also affine complete. This implies that (M, G) is a complete Riemannian manifold with Ricci curvature bounded from below by a positive constant $2L > 0$. By Myers' Theorem (see section 1.2.1) M is compact and thus an ovaloid. It follows that M is an ellipsoid (see [58], p.121).

(b) Since M is a Euclidean complete and also an affine complete affine maximal surface, by Proposition 5.2.7, M must be an elliptic paraboloid.

(c) Since the tensor norm of the Fubini-Pick tensor is equiaffinely invariant, we replace $\|B\|_G^2$ by this tensor norm; then a similar argument shows that the tensor norms of the Fubini-Pick tensor and the affine Weingarten tensor and the tensor norms of their k-th covariant derivatives all are bounded. This completes the proof of Theorem 6.1.2. ■

6.2 Hypersurfaces with Negative Constant Mean Curvature

In affine hypersurface theory, many geometric problems can be reduced to the study of a higher order PDE; for recent work see [53], [54], [57], [58], [59], [61], [62], [65], [91], [94], [102], [103], ect.

It is an interesting and important problem to classify all locally strongly convex, complete affine hypersurfaces with constant affine mean curvature. In this direction, in [53], the authors proved Theorems 6.1.1 and 6.1.2 (see section 6.1), where the first one, for surfaces with constant affine mean curvature, clarifies the relation between Euclidean and affine completeness; this result generalizes a known result for affine hyperspheres (see section 4.2.2).

Trivially, every hyperbolic affine hypersphere has constant affine mean curvature $L < 0$. As the class of complete hyperbolic affine hyperspheres is very large, there are many complete affine hypersurfaces with constant affine mean curvature $L < 0$. Recall that there is no explicit classification of all hyperbolic affine hyperspheres so far; see e.g. [41], [43].

We raise the following
Problems 6.2.1.

(i) *Is there a complete affine hypersurface with constant affine mean curvature $L < 0$ that is not an affine hypersphere?*
(ii) *If such a hypersurface in* (i) *exists, give an explicit representation.*
(iii) *Classify all affine hypersurfaces with constant affine mean curvature $L < 0$.*

Let us recall the rough classification of complete hyperbolic affine hyperspheres and the construction of Euclidean complete affine hypersurfaces with constant affine Gauß-Kronecker curvature. The classification of complete hyperbolic affine hyperspheres is reduced to the study of the following boundary value problem, where $\Omega \subset \mathbb{R}^n$ is a bounded convex domain and $L < 0$ is a real constant:

$$\det(u_{ij}) = (Lu)^{-n-2} \quad \text{in } \Omega, \tag{6.2.1}$$

$$u = 0 \qquad \text{on } \partial\Omega. \tag{6.2.2}$$

This boundary value problem has a smooth and strictly convex solution $u(\xi_1, .., \xi_n)$ in Ω. The following theorem is well-known (see [25], [58]).

Theorem 6.2.2. **(1)** *Let $\Omega \subset \mathbb{R}^n$ be a bounded convex domain and $L < 0$ be a constant, then there is a unique solution $u \in C^\infty(\Omega) \bigcap C^0(\bar{\Omega})$ of the boundary value problem (6.2.1)-(6.2.2). Put*

$$M = \{\, (x, f(x)) \,\},$$

where $f = f(x)$ is the Legendre transformation function of $u = u(\xi)$, then M is a complete hyperbolic affine hypersphere with constant affine mean curvature $L < 0$.

(2) *Every complete hyperbolic affine hypersphere can be obtained in this way.*

In [57], the authors studied the construction of Euclidean complete hypersurfaces with constant affine Gauß-Kronecker curvature. The problem is reduced to the study of the following two linked boundary value problems, where φ is prescribed on the boundary $\partial\Omega$ of the bounded convex domain Ω:

$$\det(u_{ij}) = (-u^*)^{-n-2} \quad \text{in } \Omega, \tag{6.2.3}$$

$$u = \varphi \quad \text{on } \partial\Omega, \tag{6.2.4}$$

where u^* is the solution of the following boundary value problem

$$\det(u^*_{ij}) = (-u^*)^{-n-2} \quad \text{in } \Omega, \tag{6.2.5}$$

$$u^* = 0 \quad \text{on } \partial\Omega. \tag{6.2.6}$$

They proved the following:

Theorem 6.2.3. *There is a unique solution $u \in C^\infty(\Omega) \bigcap C^0(\bar{\Omega})$ of the boundary value problem (6.2.3) - (6.2.6). The hypersurface M, constructed from the solution u as in Theorem 6.2.2, is a Euclidean complete hypersurface with constant affine Gauß-Kronecker curvature.*

In this section we study the construction of Euclidean complete affine hypersurfaces with negative constant affine mean curvature L. In the following subsection 6.2.1 we show that the construction can be reduced to the study of the PDE

$$\sum u^{ij} V_{ij} = nL, \tag{6.2.7}$$

where here and later $V := [\det(u_{ij})]^{\frac{-1}{n+2}}$.

Theorems 6.2.2 and 6.2.3 suggest to pose the following conjecture:

Conjecture 6.2.4.

(1) *Let $0 > L \in \mathbb{R}$, and let $\Omega \subset \mathbb{R}^n$ be a bounded convex domain, φ be a smooth, strictly convex function defined in a domain containing $\bar{\Omega}$, satisfying*

$$\det(\varphi_{ij}) \leq \tfrac{d}{2},$$

where

$$d = \left(2\sqrt{n}\ diam(\Omega)\right)^{-\frac{n(n+2)}{n+1}}.$$

Then there is a solution $u \in C^\infty(\Omega) \bigcap C^0(\overline{\Omega})$ of the following boundary value problem

$$\sum u^{ij} V_{ij} = nL \qquad in \quad \Omega, \tag{6.2.8}$$

$$u = \varphi \qquad on \quad \partial\Omega, \tag{6.2.9}$$

$$where \qquad \det(u_{ij}) := V^{-n-2} \qquad in \quad \Omega, \tag{6.2.10}$$

$$V = 0 \qquad on \quad \partial\Omega. \tag{6.2.11}$$

Again, if we define $M := \{(x, f(x))\}$ *as in Theorem 6.2.2 then the hypersurface* M *is a Euclidean complete affine hypersurface with constant affine mean curvature* $L < 0$.

(2) *Every Euclidean complete affine hypersurface with negative constant affine mean curvature can be obtained in this way.*

Here we solve the first part of this conjecture, namely we consider the following boundary value problem: Let Ω be a bounded convex domain with smooth boundary, let φ, ψ be given smooth functions on $\partial\Omega$ such that ψ satisfies

$$C^{-1} \leq \psi \leq C$$

for some constant $C > 0$. Solve the problem:

$$\sum u^{ij} V_{ij} \ = -n \quad \text{in} \ \ \Omega, \tag{6.2.12}$$

$$u = \varphi \qquad \text{on} \quad \partial\Omega \tag{6.2.13}$$

$$\det(u_{ij}) := \ V^{-n-2} \qquad \text{in} \ \ \Omega, \tag{6.2.14}$$

$$V = \psi \qquad \text{on} \quad \partial\Omega. \tag{6.2.15}$$

Remark. In [94], Trudinger and Wang studied the construction of hypersurfaces with given affine mean curvature function g defined on a convex domain in the coordinate plane $\mathbb{R}^2$. In our notation, they studied the following boundary value problem:

$$\sum F^{ij} w_{ij} = g, \quad w := [\det(f_{ij})]^{-\frac{n+1}{n+2}} \qquad \text{in} \ \Omega,$$

where (F^{ij}) is the cofactor matrix of the Hessian matrix (f_{ij});

$$f = \varphi, \qquad w = \psi \qquad \text{on} \ \partial\Omega,$$

where φ and ψ are prescribed functions on the boundary. As before,

$$M = \{\, (x, f(x))\, \}$$

describes the hypersurface as a graph over Ω.

Using the method of Trudinger and Wang, one can prove:

Proposition 6.2.5. *The boundary value problem* (6.2.12) – (6.2.15) *has a unique solution* $(u, V) \in C^\infty(\Omega) \bigcap C^0(\bar{\Omega})$.

The proof follows the lines of [91], thus we omit it here.

Later we are going to apply Proposition 6.2.5 and Proposition 6.2.8 below to construct a hypersurface with constant affine mean curvature $L < 0$ from the solution

u of (6.2.12) and (6.2.15) as follows. Put $M := \{(x, f(x))\}$ as in Theorem 6.2.2, then M is an affine hypersurface with constant affine mean curvature $L < 0$.

In particular we are interested to construct Euclidean complete hypersurfaces with $L < 0$. To this end we consider the following boundary value problem: Consider the PDEs (6.2.12) - (6.2.14) and

$$V = t \quad \text{on} \quad \partial\Omega, \tag{6.2.16}$$

where Ω is a bounded convex domain with smooth boundary, $t > 0$ is a given real constant, and φ is a given smooth function prescribed on the boundary $\partial\Omega$ as in the Conjecture above.

By Proposition 6.2.5 we get a family of solutions (u_t, V_t) of the PDEs (6.2.12), (6.2.13), (6.2.14) and (6.2.16). Let $t \to 0$. We can prove that u_t converges to a smooth, strictly convex function u; from u we can construct a Euclidean complete hypersurface with affine mean curvature $L < 0$ as before.

Theorem 6.2.6. [97]. *Let $\Omega \subset \mathbb{R}^n$ be a bounded convex domain with smooth boundary, φ be a smooth strictly convex function defined in a domain containing $\bar{\Omega}$, satisfying*

$$\det(\varphi_{ij}) \leq \tfrac{d}{2},$$

where

$$d = \left(2\sqrt{n}\ diam(\Omega)\right)^{-\frac{n(n+2)}{n+1}}.$$

Then there is a function $u \in C^\infty(\Omega) \bigcap C^0(\bar{\Omega})$ such that

- *u satisfies*

$$\sum u^{ij} V_{ij} = nL, \qquad \det(u_{ij}) := V^{-n-2} \qquad in \quad \Omega, \tag{6.2.17}$$

$$u = \varphi, \quad V = 0 \qquad on \quad \partial\Omega. \tag{6.2.18}$$

- *The hypersurface M, defined by the graph of the Legendre transform function f of u, is a Euclidean complete affine hypersurface with constant affine mean curvature $L < 0$. Moreover, when n=2, M is also complete with respect to the Blaschke metric.*

Remark. For a general convex domain Ω, i.e., a domain Ω with continuous boundary, we approximate Ω by convex domains that have a smooth boundary. The following result can be easily proved:

Theorem 6.2.7. *Let $\Omega \subset \mathbb{R}^n$ be a bounded convex domain with continuous boundary, φ be a smooth strictly convex function defined in a domain containing $\bar{\Omega}$, satisfying*

$$\det(\varphi_{ij}) \leq \tfrac{d}{2},$$

where

$$d = \left(2\sqrt{n}\ diam(\Omega)\right)^{-\frac{n(n+2)}{n+1}}.$$

Then there is a function $u \in C^\infty(\Omega) \bigcap C^0(\bar{\Omega})$ such that

- *u satisfies*

$$\sum u^{ij}V_{ij} = nL, \qquad \det(u_{ij}) = V^{-n-2} \qquad in \quad \Omega, \tag{6.2.19}$$

$$\lim_{p\to\partial\Omega} u \le \varphi, \quad V = 0 \qquad on \quad \partial\Omega. \tag{6.2.20}$$

- *The hypersurface M, defined by the graph of the Legendre transform f of u, is a Euclidean complete affine hypersurface with constant affine mean curvature $L < 0$. Moreover, when n=2, M is also complete with respect to the Blaschke metric.*

The proof is contained in the sections 6.2.1 - 6.2.2.

6.2.1 *Proof of the existence of a solution*

Recall the definition of ρ from section 1.4. Denote $\rho_i := \frac{\partial\rho}{\partial\xi_i}$, $V_i := \frac{\partial V}{\partial\xi_i}$, $V_{ij} := \frac{\partial^2 V}{\partial\xi_i\partial\xi_j}$, then (2.7.8) gives

$$\Delta\left(\tfrac{1}{V}\right) = \tfrac{1}{\rho}\sum u^{ij}\left(-\tfrac{V_{ij}}{V^2} + 2\tfrac{V_iV_j}{V^3}\right) - \tfrac{2}{\rho^2}\sum u^{ij}\tfrac{V_iV_j}{V^4} = -\rho\sum u^{ij}V_{ij}.$$

On the other hand, by (2.7.2) we have

$$\Delta\left(\tfrac{1}{V}\right) = \Delta\rho = -nL\rho.$$

It follows that

$$\sum u^{ij}V_{ij} = nL;$$

as before (u^{ij}) is the inverse matrix of (u_{ij}) and $V = [\det(u_{ij})]^{-\frac{1}{n+2}}$. Thus we have

Proposition 6.2.8. *Let $x : M \to \mathbb{R}^{n+1}$ be a hypersurface given by the graph of a strictly convex C^∞-function*

$$x^{n+1} = f(x^1, ..., x^n).$$

Then M has negative constant affine mean curvature $L < 0$ if and only if the Legendre transformation function u of f satisfies the following second order nonlinear PDE system

$$\begin{cases} \sum u^{ij}V_{ij} = nL \\ \\ \det(u_{ij}) = V^{-n-2}. \end{cases} \tag{6.2.21}$$

We assume that Ω is a bounded convex domain with smooth boundary. For the bounded convex domain with continuous boundary we use convex domains with smooth boundaries to approximate the given domain. As our estimates are uniform, the result follows. Without loss of generality we assume that $L = -1$.

The proof of Proposition 6.2.5 is the same as in [91], as already stated we omit it here. Now we consider the boundary value problem stated in (6.2.12) - (6.2.14) and (6.2.16). By Proposition 6.2.5 we get a one-parameter family of solutions $(u^{(t)}, V^{(t)})$

of the system. Let $t \to 0$. We are going to prove that there exists a limit function $u \in C^\infty(\Omega) \bigcap C^0(\bar{\Omega})$, which satisfies the equations (6.2.12) - (6.2.14). To this end we give uniform estimates for $|u^{(t)}|$, $V^{(t)}$ and $\rho^{(t)}$. To simplify the notation in the following estimates we use u, V, ρ instead of $u^{(t)}$, $V^{(t)}$, $\rho^{(t)}$, respectively.

(1) Estimate for $V^{(t)}$.

For any $t \in (0, 1]$, consider the function

$$F := \ln V + m \sum (\xi_i)^2$$

defined on Ω, where m is a positive constant to be determined later. If F attains its maximum at the boundary of $\partial\Omega$, it is easy to see that V has a uniform upper bound. We assume that F attains its maximum at an interior point $\xi_0 \in \Omega$, then at ξ_0, we have

$$0 = F_i = \tfrac{V_i}{V} + 2m\xi_i,$$

$$0 \geq \sum u^{ij} F_{ij} = -\tfrac{n}{V} - \sum \tfrac{u^{ij} V_i V_j}{V^2} + 2m \sum u^{ii}.$$

Without loss of generality, we may assume that the matrix (u_{ij}) is diagonal at ξ_0, then we have

$$0 \geq -\tfrac{n}{V} - 4m^2 \sum \xi_i^2 u^{ii} + 2m \sum u^{ii}.$$

We choose $m = \frac{1}{4d^2}$, where $d := \operatorname{diam}(\Omega)$. Then

$$0 \geq -\tfrac{n}{V} + m \sum u^{ii},$$

i.e.,

$$V \sum u^{ii} \leq \tfrac{n}{m}.$$

Thus, at ξ_0,

$$\sum u^{ii} = \sum \tfrac{1}{u_{ii}} \geq \frac{n}{(\Pi u_{ii})^{\frac{1}{n}}} = nV^{\frac{n+2}{n}}.$$

It follows that

$$V \leq \left(2\sqrt{n}\ \operatorname{diam}(\Omega)\right)^{\frac{n}{n+1}}. \tag{6.2.22}$$

Obviously, (6.2.22) holds everywhere in Ω.

(2) Estimate for ρ.

By the convexity of u, for any point $p \in \Omega$, the graph of u lies above the tangent plane of u at p. Subtracting a linear function we may assume that

$$u(p) = 0, \quad \operatorname{grad} u(p) = 0, \quad u(\xi) \geq 0, \quad \forall \xi \in \Omega.$$

Consider the function

$$F := \exp\left\{\tfrac{-m}{C-u}\right\} \rho,$$

defined on the *section* $S_u(p,C) = \{\xi \mid u(\xi) < C\}$, where $\rho := [\det(u_{ij})]^{\frac{1}{n+2}}$, and $m > 0$ is a constant to be determined later. It is easy to see that F attains its maximum at an interior point p^* of the domain $S_u(p,C)$. Around p^* choose a local orthonormal frame field $\{e_1, ..., e_n\}$ with respect to the Blaschke metric. Denote by "," covariant derivation with respect to the Levi-Civita connection. Then, at p^*, we have

$$-\tfrac{mu_{,i}}{(C-u)^2} + \tfrac{\rho_{,i}}{\rho} = 0, \tag{6.2.23}$$

$$-\tfrac{2m\sum(u_{,i})^2}{(C-u)^3} - \tfrac{m\Delta u}{(C-u)^2} - \tfrac{\sum(\rho_{,i})^2}{\rho^2} + \tfrac{\Delta\rho}{\rho} \leq 0. \tag{6.2.24}$$

By (2.7.8) a direct calculation gives

$$\Delta u = \tfrac{n}{\rho} - 2\sum \tfrac{\rho_{,i}}{\rho} u_{,i}. \tag{6.2.25}$$

We insert (6.2.23), (6.2.25) and (2.7.2) into (6.2.24) and get

$$-\tfrac{nm}{\rho(C-u)^2} + \left[1 - \tfrac{2(C-u)}{m}\right] \tfrac{1}{\rho^2} \sum(\rho_{,i})^2 + n \leq 0.$$

We choose $m = 2C$. Then

$$\rho \leq \tfrac{2C}{(C-u)^2}.$$

It follows that

$$\exp\left\{\tfrac{-2C}{C-u}\right\}\rho \leq \exp\left\{\tfrac{-2C}{C-u}\right\} \tfrac{2C}{(C-u)^2} \leq \tfrac{a}{C}, \tag{6.2.26}$$

where a is a universal constant. (6.2.26) holds at p^*, where F attains its maximum. Therefore (6.2.26) holds everywhere in the *section* $S_u(p,C)$.

(3) Estimate for $|u^{(t)}|$.

Adding a constant we may assume that $\min_\Omega\{\varphi\} = 0$. Let (u, V) be a solution of the system (6.2.12) - (6.2.14) and (6.2.16). Then

$$\begin{cases} u^{ij}(u+V)_{ij} = 0 & \text{in } \Omega \\ u + V = \varphi + t > 0 & \text{on } \partial\Omega. \end{cases}$$

The maximum principle implies

$$u \geq -V \geq -\left(2\sqrt{n}\ \mathrm{diam}(\Omega)\right)^{\frac{n}{n+1}},$$

then

$$|u| \leq \max_\Omega\{\varphi\} - \min_\Omega\{\varphi\} + \left(2\sqrt{n}\ \mathrm{diam}(\Omega)\right)^{\frac{n}{n+1}} + 1,$$

where we used the estimate (6.2.22).

The estimate **(3)** implies that the expression $|u^{(t)}|$ is uniformly bounded for all t. For any compact set $D \subset \Omega$ also the gradient $\|\mathrm{grad}\ u^{(t)}\|$ is uniformly bounded on D. It follows that there is a convex function u, defined on Ω, such that, for any compact set $D \subset \Omega$, there is a subsequence $u^{(t_i)}$ converging uniformly to u on D. We need to prove that u is smooth and strictly convex. We simplify the notation and write $u^{(i)} := u^{(t_i)}$. By a standard result ([17], p.369) we can find a smooth convex function φ' such that

$$\det(\varphi'_{kj}) = \tfrac{3}{4}\, d, \quad \varphi'|_{\partial\Omega} = \varphi|_{\partial\Omega},$$

where

$$d := \left(2\sqrt{n}\ \mathrm{diam}(\Omega)\right)^{-\frac{n(n+2)}{n+1}}.$$

(6.2.22) gives a uniform estimate

$$\det(u^{(i)}_{kj}) \geq d.$$

Since $u^{(i)} = \varphi$ on $\partial\Omega$, $\det(\varphi_{ij}) \leq \frac{d}{2}$; the maximum principle implies

$$\varphi \geq \varphi' > u^{(i)} \quad \text{in} \quad \Omega.$$

It follows that

$$\varphi \geq \varphi' > u. \tag{6.2.27}$$

Let $q_o = (\xi_o, u(\xi_o)) \in M^{(\infty)}$ be an arbitrary point. By (6.2.27), there is a support hyperplane $\mathbb{H}$ of $M^{(\infty)}$ at q_o

$$\mathbb{H}: \quad \xi_{n+1} = u(\xi_o) + \mathrm{grad}\, u(\xi_o)(\xi - \xi_o),$$

and a constant $a > 0$ such that

- the set
$$S(\xi_o, a) = \{\xi \in \Omega \mid u(\xi) \leq u(\xi_o) + \mathrm{grad}\, u(\xi_o)(\xi - \xi_o) + a\}$$
is not compact;
- the set
$$S(\xi_o, a-\epsilon) = \{\xi \in \Omega \mid u(\xi) \leq u(\xi_o) + \mathrm{grad}\, u(\xi_o)(\xi - \xi_o) + (a-\epsilon)\}$$
is compact for any $\epsilon > 0$.

Then there is a sequence $\xi_o^{(i)} \to \xi_o$ such that, for any small $\epsilon > 0$, the sets

$$S_i(\xi_o^{(i)}, a-\epsilon) = \left\{\xi \in \Omega \mid u^{(i)}(\xi) \leq u^{(i)}(\xi_o^{(i)}) + \mathrm{grad}\, u^{(i)}(\xi_o^{(i)})(\xi - \xi_o^{(i)}) + (a-\epsilon)\right\}$$

are also compact for i large enough. We use the estimates **(1)** and **(2)** to conclude that the expressions $\det(u^{(i)}_{kl})$ are uniformly bounded, both from above and from below. Finally we apply the Caffarelli-Gutierrez theory (see [15]) and the standard Caffarelli-Schauder estimate to conclude that u is a smooth function in $S(\xi_o, a)$ and that u satisfies the system (6.2.12) - (6.2.14). As ξ_o is arbitrary, u is smooth and strictly convex.

Now we prove

$$\lim_{p \to \partial\Omega} u = \varphi.$$

Since u is smooth and strictly convex in Ω, the limit of the left hand side exists. Denote it by φ'. Obviously, $\varphi' \leq \varphi$. We are going to prove that $\varphi' = \varphi$. Assume that there is a point $\bar{\xi} \in \partial\Omega$ such that $\varphi'(\bar{\xi}) < \varphi(\bar{\xi})$. Without loss of generality we may assume that

- $\varphi(\bar{\xi}) = 1, \quad \varphi'(\bar{\xi}) = 0,$
- $\bar{\xi} = (0, ..., 0)$, and the equation of the tangent hyperplane of $\partial\Omega$ at $\bar{\xi}$ is $\xi_1 = 0$, and $\Omega \subset \{\xi_1 > 0\}$,
- for $\epsilon > 0$ sufficiently small we have $\varphi' < \frac{1}{10}$ on $\{\xi_1 \leq \epsilon\} \bigcap \partial\Omega$.

We construct a function $\tilde{u}$

$$\tilde{u} = 2u - b\xi_1 + \tfrac{1}{2}.$$

We choose b sufficiently large and have $\tilde{u} + \frac{1}{3} \leq u^{(i)}$ on $\partial\Delta'$, where

$$\Delta' := \{\xi \in \Omega \mid \xi_1 \leq \epsilon\}.$$

For any positive real $\delta > 0$, let $D_\delta = \{\xi \in \Delta' \mid \operatorname{dist}(\xi, \partial\Delta') \geq \delta\}$. Then, for δ small enough and i large enough, we have

$$\tilde{u} < u^{(i)} \quad \text{on } \partial D_\delta,$$

$$\det(\tilde{u}_{kl}) > \det(u^{(i)}_{kl}).$$

It follows that $u^{(i)} \geq \tilde{u}$ on $\overline{D}_\delta$. For $\delta \to 0$, $i \to \infty$ we get $\tilde{u} \leq u$ in Δ'. But $\tilde{u}(\bar{\xi}) = \frac{1}{2} > \varphi'(\bar{\xi})$, and both, $\tilde{u}$ and u, are smooth in the interior of Δ'; this gives a contradiction. Thus $\varphi' = \varphi$ on $\partial\Omega$.

Put

$$x^i := \frac{\partial u}{\partial \xi_i}, \quad \text{and} \quad f(x^1, ..., x^n) := \sum \xi_i \frac{\partial u}{\partial \xi_i} - u(\xi_1, ..., \xi_n),$$

$$M = \{(x, f(x))\},$$

then M is an affine hypersurface with constant affine mean curvature -1.

6.2.2 *Proof of the Euclidean completeness*

Next we prove that M is Euclidean complete. If we prove that $\|\text{grad } u\|_E \to \infty$ when the point tends to the boundary $\partial\Omega$ then it is easy to see that M is Euclidean complete. First we use a method of [57] to show that, for any point $\bar{\xi} \in \partial\Omega$, we can find an affine transformation such that $u^{(t)}(\bar{\xi}) = 0$, and $u^{(t)}(\xi) < 0$ for all $\xi \in \partial\Omega \backslash \{\bar{\xi}\}$. To simplify the notation, in the following we omit the index t. Without loss of generality, we may assume that

$$\bar{\xi} = (0, \ldots, 0, \bar{\xi}_n), \quad \bar{\xi}_n < 0,$$

and the exterior unit normal vector of $\partial\Omega$ at $\bar{\xi}$ is $\gamma = (0, \ldots, 0, -1)$. The smooth boundary is convex, thus locally, i.e., in a neighborhood N_ε of $\bar{\xi}$ on $\partial\Omega$, the boundary $\partial\Omega$ can be expressed by

$$\xi_n = q(\xi_1, \ldots, \xi_{n-1}),$$

where $q(\xi_1, \ldots, \xi_{n-1})$ is a convex function such that

$$\frac{\partial q}{\partial \xi_i}(\bar{\xi}) = 0, \qquad \text{for} \quad i = 1, 2, \ldots, n-1.$$

Consider the function

$$F := u + b_1\xi_1 + \ldots + b_n\xi_n + d,$$

where b_i, d are constants to be determined later.

Claim. We aim to choose b_i, d such that F attains its maximum on $\partial\Omega$ at $\bar{\xi}$, and

$$F(\bar{\xi}) = 0. \tag{6.2.28}$$

To do this, we take

$$\tfrac{\partial F}{\partial \xi_i}(\bar{\xi}) = 0, \quad \text{for} \quad i = 1, 2, \ldots, n-1, \tag{6.2.29}$$

and (6.2.28), (6.2.29) give

$$b_n = \tfrac{\varphi(\bar{\xi})+d}{-\bar{\xi}_n}, \tag{6.2.30}$$

$$b_i = -\tfrac{\partial \varphi}{\partial \xi_i}(\bar{\xi}), \quad \text{for} \quad i = 1, 2, \ldots, n-1.$$

Moreover, at $\bar{\xi}$, we need

$$0 \geq \tfrac{\partial^2 F}{\partial \xi_i \partial \xi_j} = \tfrac{\partial^2 \varphi}{\partial \xi_i \partial \xi_j} + \left(\tfrac{\partial \varphi}{\partial \xi_n} + b_n\right) \tfrac{\partial^2 q}{\partial \xi_i \partial \xi_j}$$

$$= \tfrac{\partial^2 \varphi}{\partial \xi_i \partial \xi_j} + \left(\tfrac{\partial \varphi}{\partial \xi_n} + \tfrac{\varphi(\bar{\xi})+d}{-\bar{\xi}_n}\right) \tfrac{\partial^2 q}{\partial \xi_i \partial \xi_j}.$$

Denote by λ the smallest eigenvalue of the matrix $\left(\tfrac{\partial^2 q}{\partial \xi_i \partial \xi_j}\right)$, and by

$$d_1 := -\frac{\max_{N_\varepsilon}\left\{\sum_{i,j}\left|\tfrac{\partial^2 \varphi}{\partial \xi_i \partial \xi_j}\right|\right\}|\bar{\xi}_n|}{\min_{N_\varepsilon}\{\lambda\}} - \max|\varphi| - \max_{N_\varepsilon}\left|\tfrac{\partial \varphi}{\partial \xi_n}\bar{\xi}_n\right|.$$

If we choose $d < d_1$, then F attains its maximum at $\bar{\xi}$ in N_ε, and (6.2.28) holds. On the other hand, when we restrict F to $\partial\Omega$ by (6.2.13) and (6.2.30), we have

$$F = \varphi(\xi) + \sum_{i=1}^{n-1} b_i\xi_i + \left(\tfrac{\varphi(\bar{\xi})+d}{-\bar{\xi}_n}\right)\xi_n + d.$$

Denote

$$d_2 := -\max_{\partial\Omega\setminus N_\varepsilon}\left\{\tfrac{-\bar{\xi}_n}{-\bar{\xi}_n+\xi_n}\right\}\max_{\partial\Omega}\left\{|\varphi(\xi)| + \left|\sum_{i=1}^{n-1} b_i\xi_i\right| + \left|\tfrac{\varphi(\bar{\xi})}{\bar{\xi}_n}\xi_n\right|\right\}.$$

When $d < d_2$, we have $F < 0$ on $\partial\Omega\setminus N_\varepsilon$. We choose $d = 2\min\{d_1, d_2\}$, then our claim is proved. ∎

Remark. It is easy to see that F also satisfies (6.2.12), (6.2.14), and that the Euclidean gradient $\|\text{grad }(F-u)\|_E$ has a uniform upper bound. So if $\|\text{grad } F\|_E$ tends to infinity, the term $\|\text{grad } u\|_E$ tends to infinity too.

Lemma 6.2.8. *There exist a constant $\beta > 0$ and a small number $t_0 > 0$ such that, for any $t \in (0, t_0]$, the gradient of $u^{(t)}$ satisfies*

$$\|\mathit{grad}\, u^{(t)}\|_{\partial\Omega} \geq Ct^{-\beta},$$

where C is an appropriate positive constant.

Proof. Let $\bar{\xi} \in \partial\Omega$ be an arbitrary point. For a given $t \in (0,1]$, by an appropriate transformation, we may assume that $u^{(t)}(\bar{\xi}) = -t$, and $u^{(t)}(\xi) < -t$ for $\xi \in \partial\Omega \backslash \{\bar{\xi}\}$. Again, to simplify the notation, we write u, V instead of $u^{(t)}, V^{(t)}$. We have

$$\sum u^{ij}(u+V)_{ij} = 0 \qquad \text{in} \quad \Omega,$$

$$u + V \leq 0 \qquad \text{on} \quad \partial\Omega.$$

We apply the maximum principle and get

$$-V \geq u \qquad \text{on} \quad \bar{\Omega},$$

i.e.,

$$V \leq |u| \qquad \text{on} \quad \bar{\Omega}. \tag{6.2.31}$$

Let $B_\delta(\xi_0) \subset \Omega$ be a disk which is tangent to $\partial\Omega$ at $\bar{\xi}$, with radius δ and center ξ_0. Denote $a := \|\text{grad } u(\bar{\xi})\|_E$.

(1) If $a = \infty$ the lemma is proved.
(2) Now we assume that $a < \infty$. From (6.2.31), for δ sufficiently small, we have

$$V \leq |u| \leq 2\delta a + t \quad \text{on} \quad \bar{B}_\delta(\xi_0). \tag{6.2.32}$$

We construct a new function $\hat{u}$ on $B_\delta(\xi_0)$ by

$$\hat{u} := b(|\xi - \xi_0|^2 - \delta^2) - t,$$

where b is a constant to be determined later. Then

$$\tfrac{\partial \hat{u}}{\partial \gamma}(\bar{\xi}) = 2\delta b,$$
$$\det(\hat{u}_{ij}) = (2b)^n,$$

where γ is the exterior unit normal vector of $\partial\Omega$ at $\bar{\xi}$.
We take $b = \frac{1}{\delta t^\beta}$ and $\delta = t^{1-\epsilon+\beta}$, where $\beta = \frac{1-2\epsilon}{n}$, and where ϵ is a sufficiently small positive real.

Claim. We claim that $a \geq 2b\delta$.

Assume the contrary, then (6.2.32) gives

$$\det(u_{ij}) \geq (2\delta a + t)^{-n-2} > (4\delta^2 b + t)^{-n-2} > 5^{-(n+2)} t^{-(n+2)(1-\epsilon)} \quad \text{in} \quad B_\delta(\xi_0),$$

while

$$\det(\hat{u}_{ij}) = (2b)^n = \tfrac{2^n}{\delta^n t^{1-2\epsilon}} = \tfrac{2^n}{t^{n(1-\epsilon)+2(1-2\epsilon)}}\,.$$

Obviously,

$$\det(u_{ij}) \geq \det(\hat{u}_{ij}) \quad \text{in } B_\delta(\xi_0)$$

for t small. Note that $\hat{u} = -t \geq u$ on $\partial B_\delta(\xi_0)$; the maximum principle gives $\hat{u} \geq u$ on $\bar{B}_\delta(\xi_0)$. As $\hat{u}(\bar{\xi}) = -t = u(\bar{\xi})$, we have $\frac{\partial \hat{u}}{\partial \gamma}(\bar{\xi}) \leq \frac{\partial u}{\partial \gamma}(\bar{\xi})$, i.e., $a \geq 2\delta b$;

this gives a contradiction to the assumption. The claim is proved.
Therefore,

$$\|\text{grad}\, u^{(t)}\|_E \geq 2\delta b = \tfrac{2}{t^\beta},$$

and Lemma 6.2.8 is proved. ■

In the proof of Lemma 6.2.8 we obtained the inequality (6.2.31), this means that for any $t \in (0,1]$, we have $V^{(t)} \leq |u^{(t)}|$; then the limit solutions u, V also satisfy the inequality $V \leq |u|$, We apply the affine transformation used in Lemma 6.2.8; this gives $u^{(t)}(\bar{\xi}) = -t \to 0 = u(\bar{\xi})$, thus we have $V(\bar{\xi}) = 0$. On the other hand, V is invariant under the transformation in Lemma 6.2.8, and $\bar{\xi}$ is an arbitrary point of $\partial\Omega$, so $V = 0$ on $\partial\Omega$.

Denote by $f^{(t)}(x)$ and $f(x)$ the Legendre transformation functions of $u^{(t)}$ and u, respectively. For any sufficiently large number $R^\sharp > 0$, by Lemma 6.2.8, there exists a number $t_0 > 0$ such that, for $0 < t < t_0$, the function $f^{(t)}$ is defined on the disk $B_{R^\sharp}(0)$. Since the norms $\|\text{grad}\, f^{(t)}\|$ are uniformly bounded on $B_{R^\sharp}(0)$, there is a subsequence $f^{(t_i)}$ converging to f. Hence $f(x)$ is defined on $B_{R^\sharp}(0)$. As $R^\sharp$ is arbitrary, $f(x)$ is defined on $\mathbb{R}^n$, i.e., the hypersurface M constructed from u is Euclidean complete. When $n = 2$, by Theorem 6.1.1, M is also complete with respect to the Blaschke metric. ■

Bibliography

[1] M. Abreu: *Kähler geometry of toric varieties and extremal metrics*, Int. J. Math. **9** (1998), 641-651.

[2] J.A. Aledo, A. Martínez, F. Milán: *Non-removable singularities of a fourth-order nonlinear partial differential equation*, J. Diff. Equations, **247** (2009), 331-343.

[3] J.A. Aledo, A. Martínez, F. Milán: *The Cauchy problem*, J. Math. Anal. Appl. **351** (2009), 70-83.

[4] J.A. Aledo, A. Martínez, F. Milán: *Affine maximal surfaces with singularities*, Results Math. **56** (2009), 91-107.

[5] I.J. Bakelman: *Convex analysis and nonlinear geometric elliptic equations*, Springer-Verlag, Berlin (1994).

[6] A. Besse: *Einstein manifolds*, Springer, Berlin etc. (1980).

[7] T. Binder, U. Simon: *Progress in affine differential geometry–problem list and continued bibliography*, Geometry and Topology of Submanifolds X, World Scientific (2000), 1-17.

[8] T. Binder, M. Wiehe: *Invariance groups of relative normals*, Banach Center Publications, **69** (2005), 171-178.

[9] W. Blaschke: *Vorlesungen über Differentialgeometrie II. Affine Differentialgeometrie*, Springer, Berlin (1923).

[10] N. Bokan, P.B. Gilkey, U. Simon: *Geometry of differential operators on Weyl manifolds*, Proc. R. Soc. Lond. Ser. A **453**, No.1967 (1997), 2527-2536.

[11] N. Bokan, K. Nomizu, U. Simon: *Affine hypersurfaces with parallel cubic form*, Tôhoku Math. J. **42** (1990), 101-108.

[12] T. Bonnesen, W. Fenchel: *Theorie der konvexen Körper*, Springer, Berlin (1934).

[13] H. Busemann: *Convex surfaces*, Interscience Publ., New York-London (1958).

[14] L.A. Caffarelli: *Interior $W^{2,p}$ estimates for solutions of Monge-Ampère equations*, Ann. Math. **131** (1990), 135-150.

[15] L.A. Caffarelli, C.E. Gutiérrez: *Properties of the solutions of the linearized Monge-Ampère equations*, Amer. J. Math. **119** (1997), 423-464.

[16] L. Caffarelli, Y.Y. Li: *An extension to a theorem of Jörgens, Calabi, and Pogorelov*, Comm. Pure Appl. Math. **56** (2003), 549-583.

[17] L.A. Caffarelli, L. Nirenberg, J. Spruck: *The Dirichlet problem for nonlinear second order elliptic equations I. Monge-Ampère equations*, Comm. Pure Appl. Math. **37** (1984), 369-402.

[18] E. Calabi: *An extension of E. Hopf's maximum principle with an application to Riemannian geometry*, Duke Math. J. **25** (1958), 45-56.

[19] E. Calabi: *Improper affine hyperspheres of convex type and a generalization of a*

theorem by K. Jörgens, Mich. Math. J. **5** (1958), 105-126.

[20] E. Calabi: *Complete affine hypersurfaces I*, Symposia Math. **10** (1972), 19-38.

[21] E. Calabi: *Hypersurfaces with maximal affinely invariant area*, Amer. J. Math. **104** (1982), 91-126.

[22] E. Calabi: *Convex affine maximal surfaces*, Results Math. **13** (1988), 209-223.

[23] J. Cheeger, D.G. Ebin: *Comparison theorems in Riemannian geometry*, North-Holland Publishing Company (1975).

[24] G. Chen, L. Sheng: *A Bernstein property of a class of fourth order complex partial differential equations*, Results Math. to appear.

[25] S.Y. Cheng, S.T. Yau: *Complete affine hyperspheres, Part I. The completeness of affine metrics*, Comm. Pure Appl. Math. **39** (1986), 839-866.

[26] S.S. Chern: *The mathematical works of Wilhelm Blaschke*, Abh. Math. Sem. Hamburg **39** (1973), 1-9.

[27] S.S. Chern: *Affine minimal hypersurfaces*, Proc. US-Japan Seminar on Minimal Submanifolds and Geodesics, Kaigai, Tokyo (1978), 17-30.

[28] F. Dillen, K. Nomizu, L, Vrancken: *Conjugate connections and Radon's theorem in affine differential geometry*, Monatshefte Math. **109** (1990), 221-235.

[29] S.K. Donaldson: *Scalar curvatures and projective embeddings*, I, J. Diff. Geom. **59(3)** (2001), 479-522.

[30] S.K. Donaldson: *Scalar curvature and stability of toric varieties*, J. Diff. Geom. **62** (2002), 289-349.

[31] S.K. Donaldson: *Interior estimates for solutions of Abreu's equation*, arXiv: math. DG/0407486.

[32] H. Flanders: *Local theory of affine hypersurfaces*, J. Analyse Math. **15** (1965), 353-387.

[33] S. Gallot, D. Hulin, J. Lafontaine: *Riemannian geometry*, Springer Berlin etc. (1987).

[34] J.A. Galvez, A. Martinez, P. Mira: *The space of solutions to the Hessian one equation in the finitely punctured plane*, J. Math. Pures Appl. **84** (2005), 1744-1757.

[35] D. Gilbarg, N.S. Trudinger: *Elliptic partial differential equations of second order*, 2nd ed., Springer Berlin etc. (1983).

[36] P.B. Gilkey: *Geometric properties of natural operators defined by the Riemann curvature tensor*, Singapore, World Scientific. (2001).

[37] C.E. Gutiérrez: *The Monge-Ampère equation*, Birkhäuser Boston (2001).

[38] G.H. Hardy, J.E. Littlewood, G. Pólya: *Inequalities*, 2nd ed., Cambridge University Press (1988).

[39] H. Hofer: *Pseudoholomorphic curves in symplectizations with applications to the Weinstein conjecture in dimension three*, Invent. Math. **114** (1993), 515-563.

[40] Z. Hu, H. Li, U. Simon: *Schouten curvature functions on locally conformally flat Riemannian manifolds*, J. Geom. **88** (2008), 75-100.

[41] Z. Hu, H. Li, U. Simon, L. Vrancken: *On locally strongly convex affine hypersurfaces with parallel cubic form*, Part I, Diff. Geom. Appl. **27** (2009), 188-205.

[42] Z. Hu, H. Li, L. Vrancken: *Characterization of the Calabi product of hyperbolic affine hyperspheres*, Results Math. **52** (2008), 299-314.

[43] Z. Hu, H. Li, L. Vrancken: *Locally strongly convex affine hypersurfaces with parallel cubic form*, Preprint 2009.

[44] F. Jia, Y.Y. Jin: *A Bernstein property for relative extremal hypersurfaces*, J. Sichuan University, to appear.

[45] F. Jia, A.-M. Li: *Locally strongly convex hypersurfaces with constant affine mean curvature*, Diff. Geom. Appl. **22** (2005), 199-214.

[46] F. Jia, A.-M. Li: *Interior estimates for solutions of a fourth order nonlinear partial differential equation*, Diff. Geom. Appl. **25** (2007), 433-451.

[47] F. Jia, A.-M. Li: *Complete affine Kähler manifolds*, Preprint Sichuan University 2007.
[48] J. Jost, Y.L. Xin: *Some aspects of the global geometry of entire space-like submanifold*, Results Math. **40** (2001), 233-245.
[49] K. Jörgens: *Über die Lösungen der Differentialgleichung* $rt - s^2 = 1$, Math. Ann. **127** (1954), 130-134.
[50] S. Kobayashi, K. Nomizu: *Foundations of differential geometry*, vol. **I**, Interscience Publ., New York, London (1963).
[51] S. Kobayashi, K. Nomizu: *Foundations of differential geometry*, vol. **II**, Interscience Publ., New York, London, Sydney (1969).
[52] A.-M. Li, F. Jia: *The Calabi conjecture on affine maximal surfaces*, Results Math. **40** (2001), 256-272.
[53] A.-M. Li, F. Jia: *Euclidean complete affine surfaces with constant affine mean curvature*, Ann. Global Anal. Geom. **23** (2003), 283-304.
[54] A.-M. Li, F. Jia: *A Bernstein property of affine maximal hypersurfaces*, Ann. Global Anal. Geom. **23** (2003), 359-372.
[55] A.-M. Li, F. Jia: *A Bernstein property of some fourth order partial differential equations*, Results Math. **56** (2009), 109-139.
[56] A.-M. Li, H.-L. Liu, A. Schwenk-Schellschmidt, U. Simon, C.P. Wang: *Cubic form methods and relative Tchebychev hypersurfaces*, Geom. Dedicata, **66** (1997), 203-221.
[57] A.-M. Li, U. Simon, B.H. Chen: *A two-step Monge-Ampère procedure for solving a fourth order PDE for affine hypersurfaces with constant curvature*, J. Reine Angew. Math. **487** (1997), 179-200.
[58] A.-M. Li, U. Simon, G. Zhao: *Global affine differential geometry of hypersurfaces*, De Gruyter Expositions in Mathematics **11**, Berlin, Walter de Gruyter (1993), ISBN: 3-11-012769-5.
[59] A.-M. Li, U. Simon, G. Zhao: *Hypersurfaces with prescribed affine Gauß-Kronecker curvature*, Geom. Dedicata, **81** (2000), 141-166.
[60] A.-M. Li, R. Xu: *A generalization of Jörgens-Calabi-Pogorelov theorem*, J. Sichuan University, **44(5)** (2007), 1151-1152.
[61] A.-M. Li, R. Xu: *A rigidity theorem for affine Kähler-Ricci flat graph*, Results Math. **56** (2009), 141-164.
[62] A.-M. Li, R. Xu: *A cubic form differential inequality with applications to affine Kähler Ricci flat manifolds*, Results Math. **54** (2009), 329-340.
[63] A.-M. Li, G. Zhao: *Projective Blaschke manifolds*, Acta Math. Sinica, English Series **24** (2008), 1433-1448.
[64] H. Li: *A sextic holomorphic form of affine surfaces with constant affine mean curvature*, Arch. Math. (Basel) **82(3)** (2004), 263-272.
[65] H. Li: *Variational problems and PDEs in affine differential geometry*, PDEs, submanifolds and affine differential geometry, 9-41, Banach Center Publ., 69, Polish Acad. Sci., Warsaw, 2005.
[66] H.-L. Liu, U. Simon, C. P. Wang: *Conformal structure in affine geometry: complete Tchebychev hypersurfaces*, Abh. Math. Sem. Univ. Hamburg **66** (1996), 249-262.
[67] H.-L. Liu, U. Simon, C.P. Wang: *Codazzi tensors and the topology of surfaces*, Ann. Global Anal. Geom. **16** (1998), 189-202.
[68] H.-L. Liu, U. Simon, C.P. Wang: *Higher order Codazzi tensors on conformally flat spaces*, Beitr. Algebra Geom. **39** (1998), 329-348.
[69] A. Martinez: *Improper affine maps*, Math. Z. **249** (2005), 755-766.
[70] A. Martinez, F. Milan: *On the affine Bernstein problem*, Geom. Dedicata, **37** (1991), 295-302.
[71] J.A. McCoy: *A Bernstein property of solutions to a class of prescribed affine mean*

curvature equations, Ann. Global Anal Geom. **32** (2007), 147-165.

[72] K. Nomizu: *On completeness in affine differential geometry*, Geom. Dedicata, **20** (1986), 43-49.

[73] K. Nomizu, T. Sasaki: *Affine differential geometry. Geometry of affine immersions*, Cambridge Tracts in Mathematics, **111**, Cambridge University Press (1994), ISBN: 0-521-44177-3.

[74] B. Opozda: *Some relations between Riemannian and affine geometry*, Geom. Dedicata, **47** (1993), 225-236.

[75] T. Pavlista: *Geometrische Abschätzungen kleiner Eigenwerte des Laplaceoperators*, Thesis Dr. rer. nat., FB Math. TU Berlin (1984).

[76] A.V. Pogorelov: *On the improper convex affine hyperspheres*, Geom. Dedicata, **1** (1972), 33-46.

[77] A.V. Pogorelov: *The Minkowski multidimensional problem* (in Russian), Nauka, Moscow, 1975. English translation: John Wiley & Sons, New York, Toronto, London (1978).

[78] M.H. Protter, H.F. Weinberger: *Maximum principles in differential equations*, Springer New York (1984).

[79] L.A. Santalo: *La formula de Steiner para superficies paralelas en geometria afin*, Univ. Nac. Tucuman Rev., Ser. A **13** (1960), 194-208.

[80] R. Schneider: *Translations- und Ähnlichkeitssätze für Eihyperflächen*, Diploma Thesis Math. Institut Univ. Frankfurt/Main (1964).

[81] R. Schneider: *Zur affinen Differentialgeometrie im Großen I*, Math. Z. **101** (1967), 375-406.

[82] R. Schneider: *Zur affinen Differentialgeometrie im Großen II: Über eine Abschätzung der Pickschen Invariante auf Affinsphären*, Math. Z. **102** (1967), 1-8.

[83] R. Schoen, S.T. Yau: *Lectures on differential geometry*, International Press, Cambridge (1994).

[84] U. Simon: *The Pick invariant in equiaffine differential geometry*, Abh. Math. Semin. Univ. Hamb. **53** (1983), 225-228.

[85] U. Simon: *Hypersurfaces in equiaffine differential geometry*, Geom. Dedicata, **17** (1984), 157-168.

[86] U. Simon: *Affine differential geometry*, Handbook of differential geometry, Vol. **I**, 905-961, North-Holland, Amsterdam (2000).

[87] U. Simon: *Affine hypersurface theory revisited: gauge-invariant structures*, Russian Math. (Iz. VUZ) **48** (2004), 48-73.

[88] U. Simon, A. Schwenk-Schellschmidt, H. Viesel: *Introduction to the affine differential geometry of hypersurfaces*, Lecture Notes, Science University of Tokyo (1991), ISBN: 3-7983-1529-9.

[89] U. Simon, R. Xu: *Geometric modelling techniques for the solution of certain Monge-Ampère equations*, Preprint TU Berlin 2009.

[90] H. Trabelsi: *Propertiétés de l'application centre d'une hypersurface affine*, These Doctorat en Math., Univ. de Valenciennes, France (2006).

[91] N.S. Trudinger, X.-J. Wang: *The Bernstein problem for affine maximal hypersurfaces*, Invent. Math. **140** (2000), 399-422.

[92] N.S. Trudinger, X.-J. Wang: *Affine complete locally convex hypersurfaces*, Invent. Math. **150** (2002), 45-60.

[93] N.S. Trudinger, X.-J. Wang: *The Bernstein-Jörgens theorem for a fourth order partial differential equation*, J. Partial Diff. Equations, **15** (2002), 78-88.

[94] N.S. Trudinger, X.-J. Wang: *The affine Plateau problem*, J. Amer. Math. Soc. **18** (2005), 253-289.

[95] L. Vrancken: *The Magid-Ryan conjecture for equiaffine hyperspheres with constant sectional curvature*, J. Diff. Geom. **54** (2000), 99-138.

[96] L. Vrancken, A.-M. Li, U. Simon: *Affine spheres with constant affine sectional curvature*, Math. Z. **206** (1991), 651-658.

[97] B.F. Wang, A.-M. Li: *The Euclidean complete hypersurfaces with negative constant affine mean curvature*, Results Math. **52** (2008), 383-398.

[98] B.F. Wang: *The affine complete hypersurfaces of constant Gauss-Kronecker curvature*, Acta Math. Sinica, English Series **25(8)** (2009), 1353-1362.

[99] C.P. Wang: *Some examples of complete hyperbolic affine 2-spheres in* $\mathbb{R}^3$, Proc. Conf. Global Diff. Geom. Global Analysis, Berlin 1990, Lecture Notes Math. **1481** (1991), 272-280.

[100] H. Wu: *The spherical images of convex hypersurfaces*, J. Diff. Geom. **9** (1974), 279-290.

[101] M. Xiong: *Some properties of compact α-Hessian manifold*, J. Sichuan University, **45(5)** (2008), 1031-1036.

[102] R. Xu, A.-M. Li, X.X. Li: *Euclidean complete α relative extremal hypersurfaces*, J. Sichuan University, **46(5)** (2009), 1217-1223.

[103] R. Xu: *Bernstein properties for some relative parabolic affine hyperspheres*, Results Math. **52** (2008), 409-422.

[104] S.T. Yau: *Harmonic functions on complete Riemannian manifolds*, Comm. Pure. Appl. Math. **28** (1975), 201-228.

Index